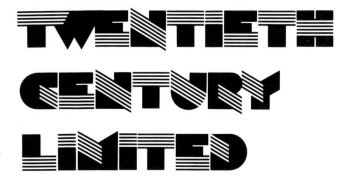

American Civilization

A Series Edited by Allen F. Davis

Gospel Hymns and Social Religion:
The Rhetoric of Nineteenth-Century Revivalism
by Sandra S. Sizer (1978)

Social Darwinism: Science and Myth in
Anglo-American Social Thought
by Robert C. Bannister (1979)

Twentieth Century Limited:
Industrial Design in America, 1925–1939
by Jeffrey L. Meikle (1979)

Charlotte Perkins Gilman:
The Making of a Radical Feminist, 1860–1896
by Mary A. Hill-Peters (1980)

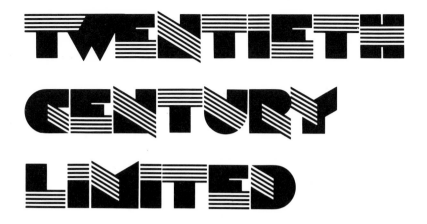

Industrial Design in America, 1925–1939

JEFFREY L. MEIKLE

Temple University Press, Philadelphia

Temple University Press, Philadelphia 19122
© 1979 by Temple University. All rights reserved
Published 1979
Printed in the United States of America

Library of Congress Cataloging in Publication Data

Meikle, Jeffrey L   1949–
   Twentieth century limited.

   (American civilization)
   Includes bibliographies and index.
   1. Design, Industrial—United States—History.
I. Title. II. Series.
TS23.M45   745.2′0973 79-17072
ISBN 0-87722-158-8

Publication of this book has been
assisted by a grant from the
Publication Program of the
National Endowment for the Humanities.

For Alice and Jason

CONTENTS

ILLUSTRATIONS

FOREWORD

Today's typical industrial designer concentrates on the limited goal of providing individual businesses and their products with public images conducive to profit. Each new trademark, appliance, fast-food outlet, or mall store is a statistically-derived package for a single corporation's version of the good life. Each design commission stands alone. In the thirties, however, when industrial design emerged as an American profession, its founders hoped to create a coherent environment for what they self-consciously referred to as "the machine age." Industrial design, they thought, would both reverse the Depression's plummeting sales and create a harmonious environment unknown since the industrial revolution. The marriage of art and industry would be renewed.

Although American industrial designers of the thirties occasionally drew on such European architects and theorists as Le Corbusier and Eric Mendelsohn, the American industrial design movement was surprisingly self-contained. The Bauhaus, which looms large in retrospect, remained little known until 1933, when its forced closing stimulated a gradual spread of its ideas to America. After Walter Gropius, Mies van der Rohe, and Marcel Breuer resettled in the United States in 1937, the Bauhaus began to have considerable impact on academics, on critics of design, and on architects, but not on commercial industrial designers, who had previously known the Bauhaus, if at all, as one of many expressions of modernism. More significant in their education was French decorative art of the twenties—now widely known as Art Deco. Even so, American designers considered streamlining, the prevalent idiom of the thirties, as a nationalistic reaction to French extravagance.

Some historians have used the term Art Deco to encompass all of the architecture and design of the period between the World Wars. David Gebhard has more perceptively distinguished between "the Zigzag Moderne of the '20s" and "the Streamlined Moderne of the '30s." In the following study I refer to the former either as "the exposition style," in order to emphasize its source in the Paris Exposition of 1925, or as "modernistic," an adjective in wide use at the time. Since the term "streamlined moderne" misleadingly implies French derivation, I prefer instead "the streamlined style" or "streamlining." "Art Deco," a term coined by Bevis Hillier in 1968, remains confusing and is most appropriate for describing a period revival of the seventies. A critic who wanted to describe in a single term such disparate designs as the Chrysler building, Le Corbusier's urban plans, and the Baby Brownie Kodak—all of which have been called examples of Art Deco—could do no better than to adopt the adjective "machine-age." Industrial designers of the thirties recognized no theory as supreme. In the American tradition of practical eclecticism, they took whatever seemed modern and transformed it for commercial use.

Many people assisted me at each stage of research and writing. First I would like to acknowledge the guidance throughout of William H. Goetzmann, whose integrative scholarship inspired whatever breadth this study possesses. Discussions with several colleagues—Suzanne Shelton Buckley, Thomas S. Hines, David Hovland, Hal Sheets, and David Tanner—helped shape my first thoughts about industrial design. For thoughtful critiques of various parts of the manuscript I am indebted to Gretchen Stevens Cochran, Malcolm Cochran, Robert M. Crunden, Theodore D. Echeverría, Robert D. Friedel, David T. Meikle, and William Stott. My editor, Kenneth Arnold, provided the encouragement needed to complete a final draft.

The staffs of several institutions assisted my research with dedication. I would like particularly to thank W. H. Crain, Jane Combs, and Ed Neal of the Hoblitzelle Theatre Arts Library, Humanities Research Center, University of Texas at Austin; Carolyn A. Davis of the George Arents Research Library at Syracuse University; Robert C. Kaufmann and David De Casseres of the Cooper-Hewitt Museum of Design, Smithsonian Institution, New York; Lois Frieman Brand of the Renwick Gallery, Smithsonian Institution, Washington; and the staff of the Inter-Library Service of the University of Texas at Austin. The following people provided valuable information, often interrupting busy schedules to do so: Edith Lutyens Bel Geddes, Leland G. Dewey, Ann Dreyfuss, John Dreyfuss, Rita Hart, Garth Huxtable, Raymond Loewy, Mrs. Eric Mendelsohn, Carl Otto, W. Dorwin Teague, Jr., and Diane Thaler. I am indebted for research grants to the American Studies Program and the Division of General and Comparative Studies of the University of Texas at Austin.

Friends and colleagues Melissa Hield, Herb Hovenkamp, Mark Smith, and Tran Qui Phiet provided much encouragement. My wife Alice gracefully remained my best critic and strongest support.

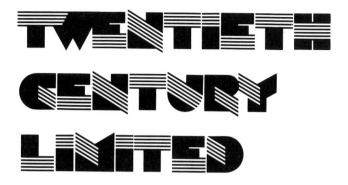

PROLOGUE

Cultural historians have come to understand the Depression of the nineteen-thirties as a time when American intellectuals and artists looked to their national past for orientation in the midst of traumatic change.[1] Historical novels filled the bestseller lists. Muralists decorated public buildings with frontier scenes. The WPA promoted regionalism by funding local histories and guidebooks. And in documenting the plight of sharecroppers and dirt farmers in hard times, photographers recalled the endurance of America's pioneers. But the common men and women of the Depression, unlike intellectuals, looked to the future for resolution of their problems.

An episode in John Steinbeck's *Grapes of Wrath*, which celebrated America's frontier spirit, ironically revealed this optimistic orientation. In flight from the Oklahoma dust bowl, the Joads have stopped for gas. Gliding by on the highway, a streamlined Lincoln Zephyr, "silvery and low," attracts the daughter's attention. Checking to see that they are alone, she asks her husband, "How'd you like to be goin' along in that?" With a sigh he replies, "Maybe—after. . . . An' if they's plenty work in California, we'll git our own car. But them . . . them kind costs as much as a good size house. I ruther have the house." Brimming with optimism, she announces, "I like to have the house *an'* one a them."[2]

When sociologist Robert Lynd surveyed Muncie, Indiana, in 1935, he concluded that even the poorest Middle Americans retained a "projective outlook on the future." Among the sources of this optimistic faith he isolated the Protestant work ethic, the frontier experience, the doctrine of evolution, and the American capitalist tradition. Most of all he stressed the "hypnotizing promise of more

and more things tomorrow" advanced by America's "machine technologies and rising standard of living."[3] Instead of seeking radical solutions to the Depression, the citizens of Muncie expected businessmen and industrialists to repair the economic system and continued to lust after goods seen in store windows and movies.

Nowhere was this faith in a future of limitless technological plenty more alive than at the Chicago Century of Progress Exposition of 1933–1934. Planned before the market crash, the exposition celebrated a machine age only a hundred years removed from pioneer days. Jagged cubist buildings, drenched with bright colors, housed numerous demonstrations of the power of science and the machine. To some visitors the exposition must have seemed an exercise in wishful thinking. Caught with nothing to celebrate, commercial exhibitors used their forum to exhort visitors to consume industry's goods in order to restore the nation's economic health. The desire for progress, they seemed to suggest, would provide its own fulfillment.

A closing ceremony held at the Great Atlantic and Pacific Tea Company exhibit echoed this theme. A crowd that had gathered to enjoy Tony Sarg's puppets and Harry Horlick's gypsy orchestra also heard a short speech by Egmont Arens. As director of the Industrial Styling Division of the Calkins and Holden advertising agency, Arens had boosted A&P's sales by redesigning its line of packaging. In his speech he proclaimed the exposition "a historic landmark"—"a turning point in our national history." According to Arens, America was moving "from an economy of scarcity into an economy of plenty," in which machines would do all work. A&P's exhibit revealed that "the distribution and the buying and preparation of food can be made into a very enjoyable thing." Spotless labor-saving appliances like those in the exhibit's experimental kitchen could "contribute to the joy of life" of America's housewives. Arens concluded by exhorting his audience to paint their kitchens the bright colors of A&P's coffee packages in order to "have a carnival in [their houses] all the year round."[4]

Some of his listeners, recalling the breadline winter of 1932, may have wondered how a bright kitchen would make up for the lack of food in it. Most, however, were caught up in his vision of a future of technological plenty, because in the exhibit this "future" was actually on display for their delight and emulation. Robert Lynd found such overriding optimism symbolized by the "strong hold on the popular imagination" of the airplane with its speed and unlimited horizons. Observing "the eagerness with which spectators throng the airport on a pleasant Sunday," he concluded that "with each new thrilling invention of this sort, the imperatives in the psychological standard of living of a portion of the population increase." Such spectacles enhanced the "desirability of making money, and lots of it," and of purchasing the new products of industrial America.[5]

Not only sociologists noted the connection between the image of flight and more mundane dreams. In March 1935 the *Bronx Home News* ran an advertisement for Sears Roebuck announcing "the New 1935 Super-Six Coldspot . . . Stunning in Its Streamlined Beauty!" This "luxurious and convenient" refrigerator, with its six cubic feet of storage space, was "new in design—modern—streamlined—arrestingly beautiful."[6] The Coldspot (figure 1) was designed by Raymond Loewy, one of a group of former craftsmen, commercial artists, stage designers, and —like Arens—advertising men engaged in creating a new commercial style with the aerodynamic image of the airplane. "From

lipsticks to locomotives" ran Loewy's slogan. In between were household appliances, clocks, radios, automobiles, packages, and retail displays—anything meant to impress consumers by its appearance. The streamlined style developed by industrial designers, as they became known, employed sweeping horizontal lines, rounded corners, and shells intended to shroud projections that might detract from an image of effortless, frictionless motion.

In a seminal study of mass production Siegfried Giedion wrote that the "anonymous" objects of everyday life "have shaken our mode of living to its very roots" because "they accumulate into forces acting upon whoever moves within the orbit of our civilization."[7] Industrial designers of the thirties recognized this fact, however uncertainly, and attempted to direct the process of influence. Industrial design itself, according to publicists, would streamline the industrial system and bring the nation out of the Depression. Products redesigned in the new style would stimulate the economy by attracting customers. Underconsumption, considered by most businessmen the nation's major economic problem, would be ended. With economic friction overcome, material progress (as well as corporate profits) would again soar like the airplane whose image inspired the movement. The streamlined idiom with its visual implications of frictionless technological progress coalesced perfectly with its economic function. More important, the American public embraced streamlining. The new style became a craze in 1934 with the emergence of the first streamliners, exhibited by the Burlington and Union Pacific railroads during the second year of the Century of Progress Exposition. In addition to serving an economic function, the style created by industrial designers in the thirties both expressed and stimulated an optimistic, often utopian mood shared by ordinary people concerned with nothing more than purchasing the latest refrigerator.

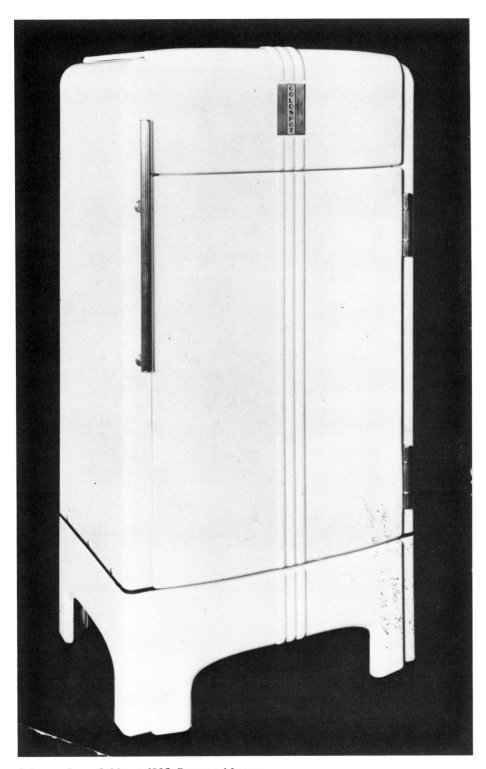

1. Loewy. Sears Coldspot, 1935. Raymond Loewy.

A Consumer Society and Its Discontents

Speed is largely a matter of mental stimulus.

—*S. L. ("Roxy") Rothafel*

Much recent social history assumes the operation of a technological imperative. According to this view, social change follows technological innovation. To Siegfried Giedion, the assembly line seemed "a symbol of the period between the two world wars."[1] Other historians have credited the automobile with stimulating prosperity in the twenties and, as its market reached saturation, with triggering the Depression of the thirties. The auto's economic role was no doubt crucial, but the ready availability of many other new manufactured products helped to transform the nation into a consumer society in the twenties. Although traditional virtues of thrift and hard work did not immediately disappear, more and more people were engaged primarily in "*buying a living*."[2]

The Flood of Goods

A number of factors contributed to the rise of a consumer economy in the twenties. The end of mass immigration in 1921 stimulated the introduction of machines to replace scarce unskilled labor. The electrification of industry made manufacturing more efficient. New techniques of factory management eliminated wasted effort by confining each worker to a single repeated task. Greater efficiency also resulted from the war, which had required standardization of parts and processes throughout entire industries, had led to modernization and new construction, and had given manufacturers experience in shifting rapidly from one line of products to another. Industrial production doubled between 1919 and 1929, while purchasing power rose by nearly a fifth. Americans who were working shorter hours for higher wages demanded a greater range of goods.[3]

By far the greatest demand was for the freedom of movement provided by an automobile. In 1910 there was one car for every 184 persons; by 1930 one for every five.[4] The automobile's phenomenal success stimulated production of other consumer goods. Its example revealed the economy of high-volume production. In addition, cost-reducing innovations sought by automakers benefited other industries. One significant automotive spin-off—the inexpensive sheet metal provided by the continuous rolling mill—transformed the home appliance industry in fortunate conjunction with lower electric rates. Other automotive developments, such as synthetic finishes, provided new appearances for old products. Lighter in weight and brighter in color, many consumer goods shared the attractions of the automobile.[5]

The acceptance of buying on credit, another automotive innovation, also fueled prosperity. Traditionally only a few products, such as pianos and sewing machines, were sold on time. After General Motors set up its own finance corporation in 1919, however, the installment plan quickly became routine for all major purchases. Noting that credit buying had become "the new gospel," Robert Lynd suggested that the desire for immediate gratification was eroding the Protestant ethic of hard work, thrift, and acceptance of hardship.[6] The present seemed capable of gratifying every desire, but Americans did not lose sight of the future. The daily increasing flow of new products engendered belief that the good life would continually grow better.

Everyone did not equally share the new goods. But, as Lynd remarked, consumers expected soon to imitate those with higher incomes.[7] Technological innovation and mass production brought former luxury items to people at lower income levels. Annual production of washing machines, for example, doubled between 1919 and 1929, and by 1931 two-thirds of the families in a typical city like Pittsburgh had one. The refrigerator, on the other hand, remained a luxury item. Not introduced until about 1916, its annual production rose from 5,000 in 1921 to 890,000 in 1929,[8] but in 1931 only 13 percent of Pittsburgh's homes had

one. But the future promised that refrigerators, like washers, would become generally available. In the meantime consumers could enjoy a range of less expensive appliances—sandwich grills, toasters, coffee pots, waffle irons, and clothing irons.

Americans desired more products that could be used or consumed quickly and easily. Canned goods, commercial bakery products, factory-rolled cigarettes, safety razors, and fountain pens all rose in sales during the twenties. They did not save much time, but they created a feeling of increased tempo. Piano sales fell sharply as people opted for public entertainments that informed them of the latest styles and products. Annual production of clothing increased by a third during the decade, indicating rapid turnover in styles. New materials like rayon found acceptance as women sought a sleek modern look to match their automobiles.

The consumption orientation spread rapidly through the United States for various reasons. The nation's population became urban-centered in the twenties, and rural areas were less isolated than before. The automobile fostered the interpenetration of rural and urban space, as city dwellers escaped to the country for Sunday drives, and farm families made shopping trips to the nearest city—whose goods made those of the general store or mail-order catalogue seem tawdry. To fight competition from urban shops, Montgomery Ward and Sears Roebuck opened retail outlets offering fashionable goods not found in their catalogues.[9] Rural visitors returned from the city to homes partially transformed by modern products representing the essence of urban living.

Just as important as greater ease of transportation in stimulating consumption was the simultaneous introduction of new ways of broadcasting information, consumption models, values, and styles. By the end of the decade Americans were flocking to twenty thousand movie theaters. Ten million homes enjoyed radios. The vast popularity of radio made people dependent on the same sources for their opinions, while the speed of communication placed a premium on novelty for its own sake. With consolidation of national networks in New York, also the headquarters of most magazine publishers, Americans both urban and rural yielded, as a government report concluded, to "mass impression" by "an all pervasive system of communication from which it is difficult to escape"[10]—a system that lubricated the wheels of industry by persuading people to buy.

The volume of advertising rose tremendously during the twenties. National periodicals ran about $25 million worth of ads in 1915. Eight years later the annual volume was up to $100 million, and in 1929 it surpassed $150 million.[11] A change in the nature of advertising was also important in boosting consumption. Before 1920 most ads simply announced availability of a given product with certain characteristics. After 1920 the industry's approach shifted to persuasion by appeal to irrational desires. Psychology, usually diluted, transformed the adman's world view. Albert T. Poffenberger, a lecturer on advertising psychology at the Columbia School of Business, advised basing ads on "the mechanism of human behavior" in "its lowest terms." An ad's "appeal" should stimulate a "reflex response" by awakening a "desire"—such as sex, success, domination, or conformity—that could be relieved through buying a given product.[12] A leading advertising executive, Earnest Elmo Calkins, defended his profession against its frequent critics. Since advertising creates "a rapid interchange of commodities and money," he argued, it "brings within easy access and at easy prices the vast number of articles . . . which make life less difficult, smoother, more restful, more efficient, and more worth while." In many cases, he observed, it "creates a demand for things that were beyond even the imagination of those who would be most benefited by them."[13]

As the major conduit of a consumption mystique derived from a faith in limitless technological progress, advertising naturally emphasized modernity, novelty, and change for its own sake. From modern art advertising executives sought ideas that would lend their illustrations a veneer of the avant-garde, no matter how drab the actual products. During the teens, autos often appeared in impressionist settings. The twenties witnessed cubist illustrations, used most often in selling clothing, cosmetics, and jewelry. By popularizing startling forms of expression from the esoteric domain of high culture, advertising itself, as a visual medium, reinforced a common perception of rapid change. Along with the automobile, labor-saving appliances, radio, movies, and new consumption-oriented styles of dressing and living, advertising both expressed and stimulated mass realization that "the tempo of life is speeded."[14]

Self-help manuals ministered to those trapped in older modes of action. Advertising executive Robert R. Updegraff included *The New American Tempo* (1929) in his "Little Library of Self-Starters." On its yellow cover marched staccato rows of blue checkmarks, one after the other, resembling stylized wings or abstractions of upward business curves. Updegraff aimed his pep talk at businessmen who watched "the sales curve of this or that item or department slowly—or perhaps abruptly—flattening out" as a result of "business competition" or "public indifference." According to Updegraff, these luckless entrepreneurs suffered failure amidst prosperity because they had ignored America's *complete change in tempo"*—including such developments as buses, tabloids, air mail, refrigeration, pale ginger ale, traffic lights, public discussion of "personal hygiene," four-wheel brakes, skyscrapers, cooperative apartments, vending machines, wirephotos, oil heaters, and "the celerity with which the nation accepted halitosis, and four out of five of us embraced the fear of pyorrhea."[15]

Faced with "the speed factor," business-men could either wait for their products to become obsolete or else emulate "a new crop of business geniuses" who had "caught the new tempo, jumped in at the right time to capitalize [on] the swing to color, the acceptance of radio, the short skirt, the craze for speed, the lure of the lurid in literature, the breaking down of prejudice against Sunday amusements, the public's discovery that it could have its 1940 luxuries today on the installment plan."[16] On the eve of the Depression, Updegraff's upbeat tone implied that he expected the future to bring ever wilder improvisations on the rhythms of the twenties. His underlying doubt, however, was revealed by his concern for businessmen who had failed in the competition for the consumer's dollar. Manufacturers began to realize the precarious balance of production and consumption. Sophisticated advertising alone no longer brought results. During the decade's final years, coached by their advertising agencies, some manufacturers began to restyle their products, making them more distinctive, modern, and competitive not only with similar goods but also with the vast assemblage of consumer products. Updegraff advised businessmen to revise their products "to fit new needs or ideas."[17] This new sales tactic soon brought a profession into being —industrial design.

The Most Expensive Art Lesson in History

One of the first to point out the importance of beauty of design in selling products—in this case automobiles—was advertising man Earnest Elmo Calkins.[18] An engaging figure who championed in voluminous writings the highest business ethics and aspirations, he was born in 1868, grew up in Galesburg, Illinois, and attended Knox College. After working as a typesetter and trade editor, Calkins gravitated to advertising, opening an agency in New York with Ralph Holden in

1901. Calkins recognized that he could make advertising more effective by improving its illustrations—usually the crude line drawings of untrained hacks—and so emphasized development of an effective art department.

As early as 1905 he suggested in his *Modern Advertising* the revolutionary idea that advertising should provide more than "the printed announcement of the merits of an article or an institution." It should also act as a "subtle, indefinable, but powerful force," creating "a demand for a given article in the minds of a great many people or arous[ing] the demand that is already there in latent form."[19] Later describing how Calkins and Holden instituted this change, he emphasized the role of art. First, skilled illustrators revealed the beauty of a product. More important, careful choice of style of illustration provided "atmosphere." Adaptation of the latest modes of the fine arts "afforded the opportunity of expressing the inexpressible, of suggesting not so much a motor car as speed, not so much a gown as style, not so much a compact as beauty."[20] Repeatedly, Calkins argued that not only did truly artistic illustrations sell products, but by acting as "a humble picture gallery for millions who never see the inside of an art museum," advertising elevated the taste of the masses, made ugly products unacceptable, and produced demand for well-designed items.[21] In addition to satisfying manufacturers by boosting sales, advertising agencies also shared responsibility for refining the material environment in which Americans lived and worked.

A Calkins and Holden campaign conducted for Pierce-Arrow around 1915 marked one of the earliest attempts to make beauty of design—rather than mechanical function—the significant factor in selling automobiles.[22] According to Calkins, the makers of the Pierce-Arrow, a luxury car, had already recognized that "people would demand finish, beauty and luxury" as soon as automobiles became reasonably similar in efficiency and performance. After hiring artists to design the Pierce-Arrow's various bodies, the company turned to Calkins and Holden for an advertising campaign. The agency created a "class atmosphere . . . conveyed . . . not so much by what was actually said, as by what it implied, and especially by the use of illustrations of superior quality reflecting the fashionable world in which the Pierce-Arrow Car lived, moved and had its being." One of the ads was framed by a dignified classical border engraved by Walter Dorwin Teague, a free-lance illustrator who had once worked in Calkins's art department and later became a leading industrial designer. Within the frame was a vaguely impressionistic oil painting. In its background stood a chauffeured Great Arrow against a backdrop of gauzy spring greenery. In front stood two well-dressed young women, one restraining a Doberman pinscher on a short leash. They appeared ready for a short spin through idyllic lanes. The accompanying copy stated that the Pierce-Arrow had reached "the ultimate degree of motor car efficiency," and its "new type of body" marked the end of "the last traditions of horse-drawn vehicles." By its studied artistry, Calkins pointed out, the advertisement portrayed an "atmosphere of luxury, comfort and smartness." Making the same point verbally in the confines of a single page would have seemed vulgar by comparison.

A three-page promotional folder prepared by Calkins and Holden could afford to be more explicit. Its double-edged title, "Leaders of Fashion," both referred to the Pierce-Arrow line and flattered potential customers. According to its copy, the Pierce-Arrow was not only "a successful machine" but also "a successful work of art, in the same way that a Sargent or a Saint Gaudens is a successful work of art." It exemplified "beauty clothing utility." Just as nature had covered the functional but ugly human skeleton with a body whose "graceful lines" suggest "strength, poise, endurance, speed, rest," so had the Pierce-Arrow company covered with an aesthetically perfect body the "engine, trans-mission, clutch—all the necessary, ugly but efficient machines that make the car what it is." The copywriter concluded by repeating that the Pierce-Arrow's body was "created out of its use and environment, created by artists."

Given Calkins's pioneering role in the use of fine artwork in advertising illustration, it follows that he would be among the first to think of designing an automobile to sell primarily by aesthetic appeal. Actually, the Great Arrow pictured in Calkins's advertisement did not differ much in appearance from one manufactured ten years before. But his agency created an impression that the car's body had been radically redesigned with beauty in mind. Implicit in the campaign was an operating philosophy later brought into the open by such industrial designers as Teague. "Leaders of Fashion" stated that the Great Arrow's body, like the streamlined housings of the thirties, covered a clutter of mechanical parts that made sense only to a mechanic. Despite the assertion that the body was "created out of its use and environment," it did not strictly honor the injunction "form follows function." The body did not express the obvious function of the Great Arrow as a transportation machine. But it did, nevertheless, express an *image* of the auto's function—an image presumed attractive to potential upper-class customers. In the thirties designers sought to represent an image of speed in their automobiles. In 1915, however, the image of the Pierce-Arrow was "class." The Pierce-Arrow's function was to provide its owner with prestige, to put one by association in a select group of people who could appreciate Sargent or Saint-Gaudens. In a general sense the function of the Pierce-Arrow was to get itself sold, and in that sense form did follow function. Later, industrial designers and their critics quibbled over whether such objects as a pencil sharpener that looked like the housing of an airplane motor exemplified functional design, but their problem concerned semantics. In 1915 the important point was that a leading advertising agency wanted the automobile sold like clothing—on the basis of style. Its design should exude the "atmosphere" of its advertising. In short, the automobile should become an advertisement for itself.

William B. Stout, a body engineer for Scripps-Booth, echoed these thoughts. Nothing then distinguished Stout from other designers, but he eventually won recognition for the Union Pacific's first streamliner and for a rear-engine auto shaped to counteract crosswinds.[23] Stating baldly as early as 1916 that "art is the science of eye-appeal," Stout argued that "if one builds into a commercial product an appeal to the eye, he establishes the first point of salesmanship, which is impression." The public no longer perceived an automobile only as a transportation machine but also as a reflection of its owner's taste and personality. Manufacturers would have to incorporate style in their designs. Stout predicted that "the car of the future" would use "art lines to suggest the action of its mechanism." Aesthetic principles as rigorous as those of mechanical engineering would yield designs whose "appeal" derived from images of "speed, power, comfort, luxury, safety and economy." According to Stout, "the automobile for to-morrow" would be "the artist's opportunity."[24]

Despite Stout's optimism it took ten years for a significant number of artists to enter the automobile industry. In 1927 the editor of *Automotive Industries* noted that only recently had designers trained in art rather than engineering begun to enter body design.[25] As long as the boxy Model T remained king of the road, concern for appearance remained limited to luxury cars like Pierce-Arrow. Gradually, other manufacturers, particularly General Motors, turned to style in order to compete with Ford. Few people realized that the automobile industry was in trouble. Owing partially to the low cost of the Model T, the auto market was approaching saturation. The year of peak increase in

total number of cars registered had passed. In 1923, 24 percent more cars were registered than in 1922. By 1926 the annual increase had dropped to 10 percent. Registrations in 1927 exceeded the previous year's by only 5 percent, while production actually fell—for the first time—by a fifth. More revealing, replacement purchases began to surpass first purchases.[26] Used-car lots overflowed with Model Ts, competing in price with new ones, while replacement customers often chose a higher-priced competitor for features not included on the Model T—longer wheelbase, standard transmission with hand shifting, six-cylinder engine, and a less functional, more stylish appearance.

Galvanized by the success of the stylish but inexpensive 1923 Chevrolet, General Motors' president, Alfred P. Sloan, Jr., began experimenting with styling throughout the firm's line. Du Pont's synthetic lacquers, introduced in 1924, made color choice important to the customer and focused attention on body design. H. Ledyard Towle, a GM color consultant, boasted that he could "make a stubby car look longer and lower" through studied use of color,[27] but inevitably someone realized that body design itself could produce the effect more convincingly. Early in 1926 Sloan hired Harley J. Earl away from a Los Angeles custom-body shop. After arriving in Detroit as a consultant to the Cadillac division, Earl created the 1927 La Salle "with a new concept in mind: that of unifying the various parts of the car from the standpoint of appearance, of rounding off sharp corners, and of lowering the silhouette." The result, "a production automobile that was as beautiful as the custom cars of the period," won Earl an appointment as director of a new Art and Color Section with ten designers and forty other employees.[28] No longer assigned to a single division, Earl and his associates contributed ideas for the complete General Motors line.

Although GM's production engineers and sales managers resisted Art and Color, Sloan kept his baby alive until it had proved itself.

In a letter of July 8, 1926, to Buick's general manager, Sloan argued that design for appearance would have "a tremendous influence on [GM's] future prosperity." "Are we," he asked, "as advanced from the standpoint of beauty of design, harmony of lines, attractiveness of color schemes and general contour of the whole piece of apparatus as we are in the soundness of workmanship and the other elements of a more mechanical nature?" [29] Repeating these themes in letters written in September 1927 to the Fisher brothers, who directed GM's body division and would be involved in retooling, Sloan boosted styling as the way to make their autos "different from competition" and "different from each other and different from year to year." [30] These letters marked the beginning of a policy of annual model changes based on elements of visual design. Planned obsolescence, orchestrated by style changes rather than technological innovations, became a mainstay of the entire automobile industry as the facts of life in a saturated market reached executive board rooms.

Sloan's innovations did not provide the most dramatic example of the imperatives of restyling. While he was formulating GM's new marketing principles, Henry Ford wrestled with his new Model A, intended to replace the car he had thought would last forever. By 1924 low-priced competitors had made noticeable inroads on Model T sales. Ford had once supposedly said, "They can have any color they want so long as it's black," [31] but after GM introduced colors in 1924, so did Ford the next year. Even that gesture did not endear the Model T to American women, who were driving more and insisting on fashionable cars. Finally submitting to the insistence of his son Edsel, Ford halted production of the Model T on May 26, 1927; by then, fifteen million had come off the assembly line. After months of work and an $18 million retooling effort, Ford produced the first Model A on October 21, and the American public hysterically greeted its public display in December. [32]

2. Ford Model T coupe, 1926. Ford Archives/Henry Ford Museum, Dearborn, Michigan.

3. Ford Model A phaeton, 1928. Courtesy of Automobile Quarterly Magazine.

With the new model Ford met his challengers technically, while maintaining an edge in pricing. More important, perhaps, its lower road clearance and longer wheelbase combined with a choice of seven body styles and eight colors to bring it in line with fashionable competitors. But a comparison of the Model A with the later Model T reveals that design changes were not radical (see figures 2 and 3). Door panel moldings disappeared. Shiny bumpers—two flat parallel strips of chrome-plated steel—appeared. The top edge of the radiator frame curved down on each side to a point at the center instead of cutting straight across. Finally, although the joint line between the body and the hood flared forward as it cut down to the mud panel, the rounding effect on the hood emphasized its separation from the body— when other automakers already stressed horizontal continuity of line. Even Edsel Ford admitted there was "nothing radical about the new car."[33] But the Model A *was* radical because Henry Ford made it. A crusty self-made inventor who had aimed at functional efficiency and economy in his machines, Ford had to give up an earlier comment that he would not "give five cents for all the art the world has produced." Instead, he parroted a public relations man by stating that "the new Ford has exceptional beauty of line and color because beauty of line and color has come to be considered, and I think rightly, a necessity in a motor car today."[34]

Shortly after the Model A's introduction, a meeting of the Society of Automotive Engineers entertained remarks from the sales promotion director of Cheney Brothers, a textile firm. Revealing "the secret of fashion and art appeal in the automobile," he warned them that automakers could satisfy demands generated by changing patterns of living only by careful analysis of style trends in all industries. Customers now assumed that all makes of cars had reached a common level of mechanical efficiency. They were shopping for models that expressed their own moods and aspirations by providing "individuality." Auto manufacturers could no longer expect to initiate trends. They could only follow them.[35] They would have to keep in mind Henry Ford if they wanted to succeed.

A few years into the Depression more than a hundred manufacturers, businessmen, and advertising men answered a questionnaire concerning the relevance of art to business. Two-thirds singled out the replacement of the Model T by the Model A in 1927 as an example of a manufacturer forced to resort to "beauty" to keep up with the modern tempo demanded by consumers.[36] Henry Ford, whose introduction of the assembly line shaped the prosperity of the twenties,

had been dragged into a world he could no longer understand. His experience, described by one observer as "the most expensive art lesson in history,"[37] served notice on other manufacturers that they could ignore demands for novelty only at their peril. The example of the Model A was significant not because the design was innovative but because it was not. The sight of an industrial giant forced to fight to keep up with more alert competitors proved instructive to lesser but equally bedeviled businessmen.

The Cash Value of Art in Industry

Even before the Depression, in the recession of 1927, businessmen had realized, as one phrased it, that "we can produce more than our people can use."[38] But few were as rational as Sloan in turning to product redesign. It is often argued that GM's annual model changes revealed cold-blooded manipulation of the public. Actually, styling resulted from a symbiotic relationship between business and consumers. Rather than manipulating the public, many manufacturers were trying to catch up with demands for novelty. One study of art in industry concluded that businessmen and designers "are not trying to force new standards on the consumer but are engaged in a rather frantic effort to ascertain what the consumer wants." The public was "setting the pace."[39] If the business sector later grew proficient in manipulating style trends, consumers in part brought the curse of planned obsolescence down on themselves.

Businessmen rarely agreed on how to introduce beauty into their products, but most recognized the factors that made the public demand it. According to one analysis, art education, sophisticated advertising, and increases in the standard of living and in leisure time had produced "a revolt against the cultural poverty which has marked American life."[40] Another observer stated that Americans were renouncing a traditional "art inferiority complex" as American civilization matured. Since most people were no longer working only for basic necessities, they now turned to cultural concerns. More to the point, the American woman had become "the great national purchasing agent." Directly involved in selecting 98 percent of all consumer goods, she applied considerations of beauty previously reserved for clothing and accessories.[41] The principle of the "ensemble"—of harmonizing different articles of dress—began to spread from fashion to more conservative industries as women sought to coordinate elements of their homes. Answering demand, for example, the Ruberoid Company offered roofing materials in a hundred designs and colors, and other businesses provided towels and soaps to match the new colored bathroom fixtures.[42]

For a while color seemed to answer the demand for beauty. Once it was introduced by the auto industry, other manufacturers sought to stimulate sales in the same way. A change in finishes did not require expensive retooling. In addition, by offering a single product in various colors, a manufacturer created a whole line of products and gave consumers opportunity to exercise choice without resorting to products of competitors. Around 1926 the market was flooded with colorful products—purple bathroom fixtures, red cookware, yellow and blue gasolines, bright plastic handles on appliances, and enameled furniture. "The Anglo-Saxon is released from chromatic inhibitions," proclaimed a headline in *Fortune*, while the more mundane *Plastics and Molded Products* asserted merely that "color sells!"[43]

Not all businessmen found a pot of gold at the end of the rainbow. Addressing an American Management Association convention, a Smith and Corona executive boasted that typewriters with color finishes made up three-fourths of sales only three years after their introduction in 1926.[44] But he did not reveal color's impact on sales as a whole.

It was true that "style, art and color" had "played a very important role in the merchandising of Corona typewriters," but competitors also turned to bright finishes. Once consumers embraced new synthetic finishes and plastics, color's role was purely negative. Without color a company would go under. With it one would survive with the rest of the pack. As promising as the color boom seemed at first, it merely complicated matters by making color a necessity and further fueling the public's demand for novelty.

American businessmen had difficulty responding to short-term fads. The opening of Tutankhamun's tomb in 1922 sparked a clamor for jewelry and clothing with Egyptian motifs. By the time a fashion designer sent to Egypt by Cheney Brothers had returned, the fad was over.[45] American manufacturers felt victimized by a merchandising system that worked well with unchanging staples but proved inadequate for coping with the tempo of the twenties. In general, manufacturers had no direct outlet to the public. They relied on merchants who carried multiple lines of the same kinds of goods and shifted from one supplier to another as the market demanded. When buying habits began to change early in the decade, larger firms invested in national advertising to stimulate habitual demand for their brands. Often such advertising merely boosted demand for a general type of merchandise without tying it to a specific brand. Merchants, especially large department stores, were in direct contact with consumers, knew what was selling, and could find somewhere a manufacturer to supply a given color or style in immediate response to demand. Manufacturers, on the other hand, had no way of predicting popular response to their current designs. Victimized by a time-lag of anywhere from several months to a year, depending on the industry, they relied on vague guesses. At times they were bitter on the subject.[46]

Department stores made matters worse by actively inciting consumers to desire novelty. Since manufacturers took the risks, merchants could only benefit by increasing the turnover of goods in their stores. Led by New York trend-setters, the nation's department stores attempted to make styles change more frequently. Adam Gimbel, the polo-playing director of Saks Fifth Avenue, returned home in 1925 from the Exposition Internationale des Arts Décoratifs et Industriels Modernes in Paris and immediately introduced its "new modernistic *décor*" in his store's show windows and interiors.[47] Lord and Taylor and Franklin Simon followed suit. Early in 1927 Macy's provided fashion-hunters with an Art-in-Trade Exposition, a "modernistic" extravaganza of foreign and domestic furnishings that attracted thousands of visitors and won praise from Robert W. de Forest, president of the Metropolitan Museum of Art's board of trustees. Department stores, he asserted, possess "a potential leadership of the utmost importance in guiding and moulding public taste and in improving the standards of design."[48] J. H. Fairclough, Jr., of Boston's Jordan Marsh furthered the argument by claiming that "department stores are the museums of today" because they "reflect good taste and act as a great educating force in the community."[49] Looking back in 1933, a member of Herbert Hoover's Research Committee on Social Trends concluded that "the aesthetic influence of the department store" had led consumers to a concern for taste in their purchases.[50] Clever publicity contributed to the mystique of the department store, but manufacturers definitely felt threatened by an institution that was popularizing "jazzy" foreign imports and thus overturning once solid public preferences. Manufacturer and merchandiser did not see eye to eye in the matter of styling.

Advertising men tended to side with retailers. William L. Day discussed the "new quandary" of "vogue and volume" in the J. Walter Thompson Agency's house organ. He began by praising manufacturers for providing Americans with mass luxuries recently undreamed of, for channeling unskilled labor into vast productive enterprises, and for unifying the nation by standardizing living habits. But despite these benefits of mass production, the industrial system yielded poor results "because of blind devotion to engineering efficiency, disregard of beauty, neglect of the human whims of the public, and disregard of those vogues which, like the impulse to migrate in primitive peoples, sweep the public away from the too efficient, too mechanical ideal set for them by mass-production." In striving for "volume," manufacturers had ignored "vogue," defined by Day as "an appeal to an ideal of beauty—or what passes for beauty—which happens to be current." Advertising, which for years had supported mass production, no longer could "shoulder the burden of its ignorance of style." To solve their problems, manufacturers would have to consult closely with retailers "in determining the character of the product."[51] American industry could no longer sustain an economy based solely on standardized mass production but would have to budget the cost of continual retooling for stylistic obsolescence.

Gradually, manufacturers realized that guesswork would not enable them to meet the public's demand for changes in style. To put product redesign on a more rational basis, they began borrowing techniques from their uncertain friends, the department stores. Retail executives routinely monitored changing buying habits by keeping records of sales and conducting market surveys to determine what consumers wanted. Manufacturers too began gathering statistics and interviewing consumers in order to forecast style trends. Some hoped to reach "beyond mere forecasting of style and fashion trends to the point of planned, thorough control of this aspect of modern business." Style was too crucial to "be allowed to drift with the wind." Instead, it "must be understood, steered and controlled."[52] A few executives developed supposedly scientific formulas for meeting if not controlling trends. Henry Creange's widely used "three-phase system" specified a third of a company's annual output as novelties, a third as successful designs from the previous year, and a third as staples recognized throughout an industry.[53] Despite interest in statistics and formulas, however, another department store innovation had more impact on manufacturers. To gain an intuitive reading of the public mind, retail executives had created the position of "stylist."

A department store stylist of the twenties engaged in little actual design work.[54] The position entailed monitoring style trends in all consumer industries and bringing a department store's offerings in line with those trends. According to Irwin D. Wolf of Pittsburgh's Kaufmann stores, the stylist's job was "to look at our merchandise through the eyes of our better educated, cultivated clientele." Sources of information included interviews with shoppers, fashionable magazines, and museum exhibits (in addition to sales statistics). After determining design trends that were expected to sell, the stylist provided the store's wholesale buyers guidelines for purchasing and its salesclerks suggestions for presenting new items to the public.[55] John E. Alcott of Jordan Marsh, who defined his occupation as "the interpretation of a mode of living," advised fellow stylists to study all conceivable forces impinging on consumers in order to derive a sense of the public's general mood. Wolf thought stylists should exert pressure on manufacturers through the agency of buyers, but Alcott suggested exerting direct pressure on a supplier's own designers.[56] If a manufacturer wanted to ensure sales of his product, he could simply order his designers to cooperate with a department store stylist, who would in turn direct wholesale buyers to purchase from the firm.

Before the appearance of department store stylists, most products were designed by engineering or art departments maintained by manufacturers. Temporary consultants were unknown. But manufacturers began to see advantages in seeking outside help. Since the public demanded novelty, product design demanded an originality not always found in employees who had spent fifteen years learning the traditions of an industry. And because department store stylists were experts on style trends in general, manufacturers began turning to them for direct application of the current mode. Charles Cannon, for example, hired Virginia Hamill, a stylist who had worked for Macy's, to apply "the new gospel of line and color" to his company's towels.[57]

Stimulated by the success stories of manufacturers like Cannon, in the final years of the decade others began to consult outsiders for help in refurbishing old products or creating new ones. In 1929 more than five hundred businessmen applied to the New York Art Center for "artists who could stylize their products."[58] Prompted by increased competition in a saturated market, shocked by the example of Henry Ford, and both led and pushed by advertising agents and retailers, American businessmen seemed ready to put their faith in "the cash value of art in industry."[59] A few complained about putting their profits at the mercy of long-haired dreamers. Others proclaimed styling a panacea. None knew what to expect.

Equally at sea were the commercial artists who answered the call of business by becoming industrial designers. What began as designing a camera or a fountain pen in spare time often mushroomed into a new career. Only one prophet saw what lay ahead for this new breed of commercial artist. Writing in the *Atlantic* of August 1927, adman Earnest Elmo Calkins predicted that giant corporations like United States Steel and the Pennsylvania Railroad would soon "have art directors whose work will be to style the products of these concerns in the aesthetic spirit of the age," so as to express "the beauty that already exists in the industrial world around us." Although he recognized that "the desire to sell" motivated businessmen to turn to artists, he hoped that business would foster an aesthetic renewal of the American environment. Released from the whimsical support of museums and private collectors, artists would again have true patrons as they had in medieval Europe. Business, the most powerful force in modern civilization, would stimulate production of art as inspiring as medieval cathedrals—and more meaningful to modern men and women. According to Calkins, Americans were "on the threshold of creating a new world on top of our modern industrial efficiency, a world in which it is possible through the much criticized machines to replace the beauty that the machines originally displaced."[60]

Whether industrial designers could succeed at such a massive reformation of the environment would depend not only on the dictates of business but also on their own training and aspirations. Most of them began by professing ideals similar to those of Calkins but became tangled in compromises as they tried to reconcile art with business and beauty with profits. Their solutions to this dilemma had a crucial impact on the American scene, and in more complex ways than Calkins foresaw.

Machine Aesthetics

*... the goods are but the vehicle
for the pattern.*

—*Richard F. Bach*

Art in Industry

Most artists who migrated to industrial
design from advertising, stage design, and
commercial illustration did so by 1929. Five
years earlier few observers would have
thought Americans capable of designing
mass production items in a style consciously
reflecting modernity and technological
progress. In fact, American design seemed
in a state of decline. Nothing demonstrated
this better than the nation's official refusal to
exhibit at the Paris exposition of 1925 on the
ground that American craftsmen and manufacturers produced only "reproductions,
imitations and counterfeits of ancient
styles."[1] The term "industrial design" first
appeared in 1919, but its meaning bore little
resemblance to the profession that began
evolving ten years later.[2] Usually designers
described themselves as engaged in "art in
industry," "art industries," "industrial arts,"
"art crafts," or "arts allied to architecture"—
terms covering such areas as textiles, ceramics, jewelry, furniture, lighting fixtures, hardware, wallpaper, and commercial illustration.[3] Most designers created decorative accessories for the home or person rather than
utilitarian objects, which were almost always
produced by engineers with no artistic training. Often, especially in the areas of furniture, hardware, and lighting fixtures, designers produced custom work for wealthy
patrons rather than mass production items.
They emphasized art, not industry, and
considered themselves craftsmen.

A number of institutions sustained this
conservative design establishment. In New
York the Art Center, with six galleries and
fourteen meeting rooms, housed a variety of
organizations with combined membership of
three thousand, including the Art Alliance of
America, the New York Society of Craftsmen,
the Art Directors Club, and the Society of
Illustrators. Although the Art Center embraced industrial design in the mid-thirties,
it emphasized craftsmanship in 1920. An
observer described it as "a rallying point for

all individuals and societies having the development of decorative and ornamental design at heart" and singled out its design classes in ceramics and textiles as influential. Interior decorators and suppliers belonged to the Art-in-Trades Club, and the Architectural League of New York included "painters, sculptors and designers in the crafts" in addition to licensed architects. In 1920 the League added a medal in "industrial art" to its competitions in architecture, painting, sculpture, and landscape architecture. Its annual show in 1924 included such derivative artifacts as stained glass, wrought iron in medieval patterns, and "Chinese lighting fixtures."[4]

Even more conservative were the design schools of America. According to one observer, both public schools and advanced art schools continued in the twenties to teach Art Nouveau, a style of elaborate patterns derived from plant forms, introduced in America around 1910.[5] Art Nouveau seemed far removed from the age of mass production; it harked back, with Craftsman furniture and Gothic-revival architecture, to a supposedly harmonious premachine epoch. Because most industrial art schools depended on museums for institutional support, they also relied on museum collections for inspirational models. Students learned to produce unique craft items derived from a wide range of historical styles. Only the Massachusetts School of Art, originally established to train public school art teachers, proved an exception to this conservative pattern. Royal Bailey Farnum, who became its director in 1921, decided "to make business more artistic without making it a whit less businesslike." Under his guidance the school emphasized art "that men can earn their living by." According to an instructor, "fine art" was "useless art," but "a man might design a garbage can which would be just as fine as anything in the Boston Museum."[6]

Such a statement met sympathy in few places. Among those who agreed was John Cotton Dana, a crusty Vermonter who directed Newark's city museum.[7] Dana led an industrial art museum movement that supported mass production design without completely rejecting the conservatism of art schools. Born in 1856 and educated at Dartmouth, he worked as a surveyor, studied law, and administered several municipal library systems before becoming director of Newark's new art museum in 1909. Immediately, he decided not to form a traditional art museum because Newark could not compete with the nearby Metropolitan Museum. Instead, he exhibited manufacturing processes and local industrial products—not only to instill Newark's citizens with pride but also to improve their taste by exposing them to selected examples of good design. Conceiving his museum as an educational tool, he strove to demonstrate that "beauty has no relation to price, rarity or age."[8] Capable of setting cases of minerals in a room of American oil paintings purchased by the federal government, he reached the epitome of plainness in 1915 with a large room full of bathtubs.[9] Dana struck most museum directors as eccentric, but a few smaller cities, too poor for old masters, heeded his gospel of utilitarian beauty and opened industrial art departments. His most important convert, however, was Richard F. Bach, who in 1918 began an industrial art program in the nation's most prestigious art museum, New York's Metropolitan.

Appointed to carry out "the application of arts to manufacture and practical life," a neglected provision of the museum's charter, Bach had worked as an architecture instructor and librarian at Columbia after his graduation in 1908.[10] Bach outlined his philosophy in an address to the American Federation of Arts in 1920. According to him, all museums should function as "active educational institutions, instruments of public service." The art museum should provide a "missionary service" to the "fertile virgin soil" of mass production by encouraging manufacturers to seek inspiration from its collections and by honoring tasteful commercial products.[11] In practicing this philosophy, Bach used several strategems. He himself contributed articles to design magazines, business periodicals, and the general press. His department organized conferences for manufacturers and advised individual firms on the use of museum collections. An annual exhibit displayed contemporary products whose designs reflected those of museum artifacts. Although Bach thus stimulated a historicism at odds with an emerging modern style, he did contribute to a concern for aesthetics in mass production. Despite his conservatism Bach prepared the way for the industrial design of the thirties with his attack on shoddiness and his belief in good design as a socially uplifting force.

In articles published early in the twenties Bach outlined his complaints about American industrial art and suggested ways of improving the situation. He blamed the low national level of taste on the prevalence of ugly household furnishings. To Americans of modest incomes, he asserted, no decent designs were available. At the mercy of artistically untrained manufacturers, they suffered "the curse of the average." Enforced mediocrity in their surroundings tragically contributed to an unawareness of beauty that narrowed their lives. Because an unenlightened public would buy anything, suppliers of household furnishings were morally bankrupt in operating under "the same commercial attitude of mind which might characterize a seller of wire and nails." Without "the element of taste or design," the industry became a "quack business." The moral responsibility for weeding out poor designs fell not on the consumer but on the manufacturer, who "owed" the public good designs. Only two courses of action would restore a national taste destroyed by mass distribution of ugly designs. First, he challenged "those leaders in the world of art that now confine themselves to the arts miscalled fine" to commit themselves to programs like the Metropolitan's. And second,

he called on craftsmen to turn from creation of unique luxury items to the design of objects for mass production.[12]

Bach believed in the regenerative possibility of a marriage of art and business. But even though he provided an altruistic rationale for industrial design, his personal taste too long remained mired in an eclecticism inspired by the Metropolitan's collections. Reviewing Bach's 1925 exhibit of industrial art, *The Architectural Record* criticized his preoccupation with historical styles and asserted that "the time has arrived when skilful imitations of the antique begin to pall, and the spirit of emancipation in artistic expression would be more than welcome."[13] That summer the Paris exposition had provided a glimpse of a thoroughly modern style, one that seemed perfect for unifying all material artifacts—from consumer products to architecture—in a holistic expression of the machine-age tempo.

A Mad and Colorful Conglomeration

Squeezed among architectural relics of France's past glory, the pavilions of the Exposition Internationale des Arts Décoratifs et Industriels Modernes (figure 4) seemed to critic Helen Appleton Read "a cubist dream city or the projection of a possible city in Mars, arisen over night in the heart of Paris."[14] Another American journalist, overcome by bright colors and unorthodox angular forms, reported that "you feel, somehow, as if in a kaleidoscope."[15] More discerning, two faculty members of Indiana University noted that French architects and designers had discarded "the entwining algae, gigantic vermicelli, and contorted medusas" of Art Nouveau "for the inorganic figures of geometry and geology—bodies bounded by plane surfaces." Some enthusiasts, they concluded, sought in this new style of "simplicity and scientific technique" a harmonious reconciliation of "art and mechanism."[16] There were, of course,

4. The Gate of Honor, Paris, 1925. International
Studio.

skeptics. A member of the United States dele-
gation considered the exposition "the most
serious and sustained exhibition of bad taste
the world has ever seen."[17] But most observers agreed with Read that it marked "a definite break with the past" in a positive sense
by exhibiting a new style for "a time differing
so radically from any preceding one"—an
epoch that had experienced electricity,
radio, and relativity.[18]

Despite contemporary perceptions of a
new style emerging from the exposition, its
lush decadence hardly marked a beginning.
Its buildings, more restrained outside than
in, exemplified the classical symmetry common to French architecture from Versailles
to Baron Haussmann's Paris. Most novelty of
effect, including the frequently cited cubism,

derived from applied ornament. The conservative Pavilion of Elegance, praised by
Read for its simplicity, had flat abstractions
of fluted columns at its corners, an overhanging cornice, and bands of ornament
above its first-story windows. Its rooms,
however, were furnished with a cluttered
exoticism (figure 5): contrasting wall panels
of rare polished woods, zebra-skin rugs,
sculptures of otherworldly plants, and
wooden mannequins with abstracted Greek
faces and flowing gowns—all created a hot-
house effect hardly expressive of a stripped-
down machine age. Exposition furniture
seemed rooted in a dead-end mixture of Art
Nouveau, orientalism, and a tradition of
unique luxury items designed for rich
patrons. Chairs and tables followed the over-

*5. Interior of the Pavilion of Elegance, Paris,
1925.* International Studio.

stuffed Empire style in form, but coverings
of all types exhibited a riotous use of colored
weavings and prints. Ornamental motifs reflected simplification of florid organic Art
Nouveau motifs, rendered geometric and
inorganic by abstraction and stylization.
Despite this development, the only hints of
the zigzag lightning motifs and cubist designs
that became hallmarks of the exposition
style as it moved to America appeared on
novelty items like cigarette boxes, vanity
cases, and jewelry.

Why, then, did Read and other perceptive
Americans consider the exposition a harbin-
ger of a design style suited to a fast-paced
machine age? Perhaps she stated the case
correctly when she wrote that the exposition had "broken the spell of the period de-

sign and spread broadly the gospel of creating design in harmony with the times."[19]
Engaged in creating neo-Tudor suburban
residences, neo-Gothic churches and cam-
puses, period furniture, and Art Nouveau
graphics, most American architects and
craftsmen knew little of their own Frank
Lloyd Wright and next to nothing of the
Bauhaus. Cut off by the war from European
developments, Americans found the exposition of 1925 a breath of fresh air. They considered it modern because it seemed to
have sprung full-blown from nothing. In it
they discerned a coherent modern style to
replace the historical styles.

In addition, although no artifacts on display were mass-produced, being unique
objects for wealthy collectors, some re-

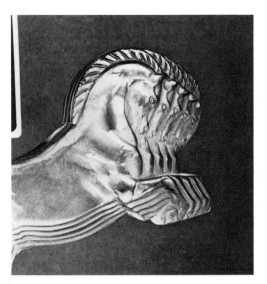

6. René Lalique. Automobile radiator ornament, c. 1925. Angelo Hornak.

flected in texture or motif the machine and its products. Edgar Brandt's ironwork, among the exposition's most admired elements, was shaped by hand, but his stylizations of organic forms like vines and peacock tails revealed unnatural precision. Characteristic repetition of identical forms throughout a particular work suggested the accuracy of repetition of mass production. Another designer, René Lalique, produced glassware so flawlessly exact as to suggest machine production. One of his designs, an auto radiator emblem, depicted in relief five identical leaping horses—calling to mind the repeatability of mass production (see figures 6 and 7). Their frozen leap into space suggested not so much the movement of living animals as an abstraction of the idea of speed.

Beyond the effects of individual designs, the rise of the *ensemblier* proved influential in America. Most French designers created complete rooms by carefully designing and orchestrating furniture, rugs, tapestries, wallpaper, and minor accessories for a unity of effect. In America, although Wright had instituted this practice, often disrupting his clients' lives, the French approach seemed innovative. The work of ensembliers was

limited to interiors for wealthy patrons, but their operating philosophy influenced industrial design as it developed in the United States. From their approach evolved the idea of creating not only a product but its factory as well, not only a stove but its house too, and conversely, not only a passenger train but also its minor elements, uniforms, menus, and matchbooks. Some American designers eventually theoretically extended this principle of total control to encompass the redesign of the entire environment. Thus the Paris exposition, dedicated to rather minor decorative arts, contributed to a vision of the world harmoniously made over by a supreme generalist, the industrial designer. But this concept germinated slowly. Most Americans who attended the exposition had concerns more immediate—and more practical.

America owed its limited participation in the exposition to Cheney Brothers, the innovative textile firm that in 1924 had introduced a line of "Ferroniere prints" based on Brandt's ironwork patterns. At their insistence, Secretary of Commerce Herbert Hoover appointed three commissioners who invited a wide range of manufacturers and artists to attend as official delegates at their own expense.[20] More than a hundred went, representing virtually every trade association and art guild. This response demonstrated widespread desire for exposure to new ideas.[21] Almost immediately, the work displayed at the exposition made its impact felt in the United States. A sampling of furniture ensembles, ceramics, glassware, metalwork, and textiles toured the Metropolitan and eight other museums early in 1926—enabling Americans who had not pilgrimaged to Paris to meet the style at first hand.[22] Simultaneously, the Metropolitan established a permanent gallery for displaying its purchases of modern decorative objects, a few dating from the early twenties but most acquired at the exposition.[23] Not until 1934 did Montgomery Ward offer for mass consumption products derived from

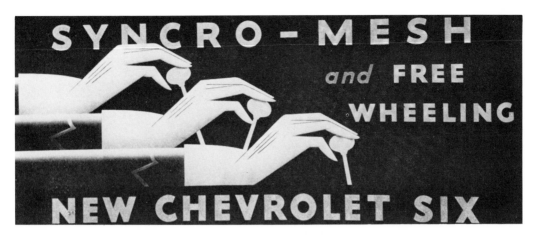

7. *James A. Kelley of Campbell-Ewald Co.*
Chevrolet billboard, 1932. Crain
Communications, Inc.

the exposition, but by 1927 most fashionable department stores, led by those of New York, had followed the museums' lead. Looking back from the thirties, *Fortune* noted that soon after the exposition had "crystallized interest in *modernique*," a plethora of "modernist backgrounds" swept Fifth Avenue window displays, consisting of the "grotesque mannequins, the cubist props, the gaga designs" that became the style's hallmarks.[24] Rejecting the traditional strewing of unrelated products against polished oak paneling, window decorators used contrasting grays, arranged in jagged angular patterns, against which a stylized mannequin modeled an accessory or two, spotlighted to emphasize shadows created by the background patterns.[25]

Even more spectacular were design exhibits mounted by stores. Macy's and Lord and Taylor outfitted "galleries" containing their selections of the best modernistic decorative art. Macy's Art-in-Trade Exposition of 1927 benefited from consultants Robert W. de Forest of the Metropolitan and John H. Finley of the New York City Arts Council. Lee Simonson's backgrounds (figure 8), panels of cork flooring material jutting out at sharp angles to one another, provided the visual effects of a cubist painting, but they

also reflected the disorienting expressionism of *The Cabinet of Dr. Caligari*, released in the United States in 1921. In addition to showing French designs, Macy's exhibited American graphics, ceramics, and silks with repetitive machinelike patterns of airplanes, skyscrapers, and dancers. These displayed more geometric angularity and fewer organic motifs than French designs, as American decorative artists moved consciously toward a machine aesthetic.[26]

Simonson himself employed a rhetoric common among designers who adapted the modernistic style for Americans. Referring to Macy's exhibit, he told a reporter that "art in this age has to be simplified if it is going to be produced on a large enough scale for most people to afford and enjoy it." Mass production would yield "a tremendous widening of the variety of articles in everyday use as generally accepted essentials of modern life."[27] Simonson's remarks exhibited an egalitarian spirit, reminiscent of Bach's exhortation to manufacturers to improve public taste or of Calkins's vision of America remade by designers. Unfortunately, the masses might look at the modish furnishings arranged by Simonson, but they could not afford them. Macy's exhibit revealed an elitism similar to that of French designers. Some

8. Lee Simonson. Displays for Macy's
Art-in-Trade Exposition, 1927. Crain
Communications, Inc.

items were unique, others were produced in limited quantities for the wealthiest consumers. Such artifacts could be produced in quantity only to the extent that hand craftsmanship would allow.

This split between a faith in the social benefit of design for mass production and the reality of custom-made luxury goods appeared glaringly in a slick *Annual of American Design 1931*. Its coeditor exulted that five years after admitting "decorative insignificance," the nation boasted "a powerful and conscious movement in American industrial design that can stand comparison favorably with the best done in Europe."[28] An ambitious production, the *Annual* contained articles by experts in various design fields illustrated with portfolios of relevant photographs. Most contributors discussed the enormous potential of intelligently directed mass production for alleviating

suffering and relieving the dullness of the artificial environment. In the lead essay, "Culture and Machine Art," Lewis Mumford defined machine production as inherently democratic. It could not be linked to Thorstein Veblen's conspicuous consumption—typical of middle-class society with its emphasis on custom craftsmanship and the house as museum.[29] Paul T. Frankl, a designer who specialized in luxury furniture, paradoxically told his colleagues that they should provide the masses with stylish but inexpensive furnishings instead of catering to "the irresponsible and the sophisticated."[30] Finally, Richard F. Bach typically noted that a frequent criticism of the machine—that its products fell to a lowest common denominator and cheapened public taste—actually should be aimed at manufacturers who employed incompetent designers.[31] Against this verbal uniformity of concern for mass production, the accompanying photographs, presumably chosen by the *Annual*'s editors, astonish one by their incongruity. Virtually all illustrated the very custom-made luxury objects being so vociferously attacked. The volume's illustrations thus portrayed designs with no connection to the mass production rhetoric filling its text.

Several factors contributed to this divergence of practices from ideals. First, native American designers educated before 1925 continued to see themselves as craftsmen, however much the Paris exposition had changed their stylistic preferences. They cared little for and knew nothing about the processes and materials of mass production. Second, along with museum and department store exhibits, practicing architects contributed equally to spreading the new modernism. Trained in the Beaux-Arts school and receptive to anything French, architects like Raymond Hood, Harvey Wiley Corbett, and Ely Jacques Kahn easily switched to the new style. Contemporaries remembered them as "leaders of a small group that labored on many committees for the promotion of modernism in the applied arts."[32] Although they made a few forays into industrial design, they tended to follow the French ensembliers by designing fixtures and interiors for their architectural projects (see figure 9). Once again the emphasis was on custom design. They followed architectural tradition by concerning themselves with individual clients rather than with manufacturers, and those clients tended to be the most avant-garde of the social elite.

Finally, a third factor contributed to the discrepancy between the ideal of renewing America through mass production and the reality of an elite craft movement. The American rage for the modernistic induced a large number of foreign designers to immigrate. This influx reached such proportions that the American Union of Decorative Artists and Craftsmen often conducted meetings in German. A list of immigrants compiled by Frankl, himself from Vienna, read

like a roll call of leading American designers of the late twenties. Kem Weber from Berlin concentrated on furniture and commercial interiors in Los Angeles. Joseph Urban of Vienna devoted himself at first to theater interiors. Winold Reiss, who had come from Karlsruhe, Germany, served as an ensemblier of large public interiors. From Switzerland came William Lescaze, whose responsibility in his architectural partnership with George Howe included the interiors of their collaborations. Most foreign designers, though steeped in prewar German expressionism, combined their earlier experience with the French style demanded by Americans.[33] Although they called themselves industrial designers, they too, like native American architects and decorative artists, primarily produced custom-made objects and unique ensembles. Their forays into mass production design were limited to novelties and personal accessories. Thus they posed little threat to a new breed of industrial designers with few inhibiting ties to the decorative art establishment. But by developing modernism in America, these immigrants helped provide a design vocabulary expressive of the tempo of America's new machine age that could be applied by others to consumer products.

Reporting from Paris in 1925, a journalist had complained that the exposition revealed "no harmony, aesthetic or supra-aesthetic— that is to say moral"—and concluded that "to have a style of its own, a period must have an inner harmony."[34] Increasingly, in the late twenties American designers found this harmony in the image of the machine with its attributes of speed, efficiency, precision, and reliability. And as for a moral dimension to design, publicists and apologists repeatedly stressed the social benefits of progress, the social harmonies possible through intelligent direction of the machine, and the fitness of an environment made over in its image. The resulting machine aesthetic provided a satisfying rationale for industrial designers who found themselves often at

9. Raymond Hood. Apartment house loggia for an exhibit on "The Architect and the Industrial Arts," held by the Metropolitan Museum of Art, 1929. Metropolitan Museum of Art.

odds with businessmen whose interest in design went no further than increases in profits.

The Era of the Machine

Architectural historians generally agree that modern architecture reached America through European theorists who admired the functionalism of American factories, grain elevators, bridges, and steel-frame construction, and who sought to formalize aesthetically what had been a matter of engineering efficiency. Until recently, as David Gebhard has noted, our understanding of the twentieth-century machine aesthetic has relied too heavily on the "handsomely designed historical construct" of Nikolaus Pevsner's *Pioneers of the Modern Movement* (1936) and Siegfried Giedion's *Space, Time and Architecture* (1941). Each described a linear history of functionalism that relegated important figures like expressionist Eric Mendelsohn or modernist Raymond Hood to a limbo of bad taste. Gebhard describes the International Style as one of several "stylistic containers" of the twenties and thirties—a critical approach that makes possible new assessments of architects who too long remained pigeonholed.[35] Le Corbusier, whose seminal *Towards a New Architecture* appeared in English in 1927, emerges as a contributor to an expressionist machine aesthetic developed by American designers to rationalize their adaptations of the exposition's modernistic style and later their own streamlined style. Since he based his vision on American industrial architecture, claiming that "THE AMERICAN ENGINEERS OVERWHELM WITH THEIR CALCULATIONS OUR EXPIRING ARCHITECTURE," they could feel that in following him they were pursuing a truly American style—a crucial consideration in the face of their sense of inferiority following the Paris exposition.[36]

According to Le Corbusier, modern men and women suffered a malaise or vague dis-ease because their environment, buildings either hundreds of years old or designed as if they were, did not correspond to modern life. The architect's first concern, he thought, was to provide "the measure of an order which we feel to be in accordance with that of our world," to create a style to replace the historical "styles." Relying on "the Engineer's Æsthetic . . . inspired by the law of Economy and governed by mathematical calculation," the architect would create an eternal style whose "platonic grandeur" would resonate on a "sounding-board" possessed innately by every person. Throughout Le Corbusier's manifesto ran a desire for stasis, for attaining perfect social equilibrium through design. He conceived of society as a machine, "profoundly *out of gear*" owing to inadequate housing that was driving the masses toward revolution. Only the architect-designer, with "a well-mapped-out scheme" for mass-produced housing, could provide "a feeling of calm, order and neatness," thus "impos[ing] discipline on the inhabitants." Le Corbusier desired a progression from the organic to the inorganic, from the natural to the artificial, from the random uncertainties of life to the reliability of the perfect machine. He espoused "the replacing of natural materials by artificial ones, of heterogeneous and doubtful materials by homogeneous and artificial ones (tried and proved in the laboratory) and by products of fixed composition." Eventually, a new environment patterned on the smooth functioning of the machine would produce a race in the image of the architect-designer— "intelligent, cold and calm," and presumably unquestioning as its members fulfilled their functions in a vast social machine.

Not until the Depression did American industrial designers heed Le Corbusier's vision of social stasis, but other elements of *Towards a New Architecture* immediately entered the thoughts of designers and architects. The craftsmen of AUDAC, writing a constitution in 1928, echoed Le Corbusier by stating their main goal as eliminating "the discrepancy between the life of the people of today and the setting in which it is lived."[37] Paul Frankl stressed a need for "a style of today and tomorrow, created by artists who understand our lives and our likes, who represent our unconscious wishes and desires." Almost quoting Le Corbusier, he declared that "modernism means style versus styles."[38] In *The Metropolis of Tomorrow* Hugh Ferriss discussed the problem of "a human population unconsciously reacting to forms which came into existence without conscious design," but he hoped for the appearance of "architects who, appreciating the influence unconsciously received, will learn consciously to direct it."[39]

Most American commentators agreed with Le Corbusier that the mathematical precision of engineering should govern the new aesthetic. In 1928 Lewis Mumford described engineering as "an exact art." According to him, "industrialism" had "produced new arts, associated with the application of precise methods and machine tools." He praised purely functional designs of automobiles, airplanes, kitchen appliances, and bathroom fixtures. He predicted, nevertheless, extension of "the machine-technique into fields of activity where the personality as a whole must be considered, and where social adaptations and psychological stresses and strains are just as important factors as tensile strength, load, or mechanical efficiency in operation."[40] As a social critic, Mumford emphasized more than most practicing designers the reform tendency of Le Corbusier's message, but they too looked to the engineer for inspiration. Frederick Kiesler declared in 1930 that "the new beauty must be based on EFFICIENCY and not on decorative cosmetics"; and Donald Deskey claimed three years later that designers and architects had rejected surface ornament after learning efficiency from engineering.[41]

Others agreed with Le Corbusier that the machine represented a Platonic ideal form and that a machine aesthetic would provide beauty innately apparent to anyone. John

Cotton Dana in 1928 characterized the development of "machine art" as "a search for technical form, which is fundamentally a form of simple lines with a minimum of surface decoration." In order "to meet the demands of machine production," the new design style would rely on "geometric simplicity, and the beauty of that kind of form depends upon mathematics and upon that fact that we can apprehend and appreciate it clearly."[42] Le Corbusier's influence appeared more certainly in Frankl's *Form and Reform*. According to him, "modernism" was a "style of reason," making "an appeal to the intelligence" and marking "a return to the Greek ideal." By "its emphasis upon simple forms, its return to mathematical axiom and the fundamentals of form," modernism became "a classical rather than a romantic beauty." Through a scientific arrangement of "line, proportion and inherent relations," he argued, "the new classicism" made "a direct appeal to the vision and the mind."[43]

No doubt Frankl took seriously his Corbusian classicism, but illustrations in his volume indicated that many designers were actually concerned with expressing subjective impressions induced by the machine. Nothing was more romantic than Frankl's own skyscraper furniture (figure 10), based on a perception of the skyscraper as the machine age's ultimate architectural symbol. A new "spirit" of "speed, compression, directness," he wrote, "*finds expression* in skyscrapers, motor-cars, airplanes, in new ocean liners, in department stores and great industrial plants." Even his chapter on new industrial materials like Vitrolite glass, Bakelite plastic, and Monel metal waxed romantic, entitled "Materia Nova." Through the "alchemy" of synthetic chemistry, "base materials" were "transmuted into marvels of beauty," considered by Frankl "*expressive* of our own age."[44]

Despite a functionalist reputation, Le Corbusier himself steered American designers toward an expressionist machine aesthetic, according to which consumer products,

10. *Paul T. Frankl. Room with a skyscraper book case, c. 1929.* International Studio.

transportation machines, buildings, and interiors would reflect attributes of the machine—speed, power, precision, machined surfaces, and impersonality. Although he argued that a house should be planned as logically as an airplane to facilitate its inhabitants' functions, his design philosophy went beyond functionalism. "The business of Architecture," he wrote, "is to establish emotional relationships by means of raw materials," to go "beyond utilitarian needs."[45] He revealed his own emotive bent in a discussion of ocean liners. In general, the accompanying photographs portrayed abstract massings of white decks and bridges, punctuated by windows and slender uprights at regular intervals, and broken horizontally by dark railings. His readers must have noticed the resemblance of these composed photographs to the buildings of the Weissenhof suburb, which in 1927 gave

birth to the International Style. Le Corbusier himself provided an example of an architect imitating forms of transportation machines —and for their expressive value, though he might have denied it. When American designers later streamlined stationary products and buildings, they followed Le Corbusier's expressive use of the machine. The new American tempo merely dictated a flashier model than the ocean liner of the European style.

Another element of Le Corbusier's manifesto appealed to American designers fresh from art-in-industry programs of Dana, Bach, and Calkins. Only "big business," asserted Le Corbusier, had the resources to prevent revolution by applying architecture. No longer a conservative social force, business had "modified its customs" to become "a healthy and moral organism." He found business marked by "intelligence" and "bold innovations" because "engineers in number fill its offices, make their calculations, practise the laws of economy to an intensive degree, and seek to harmonize two opposed factors: cheapness and good work."[46] Devoid of consideration of the profit motive, these comments expose wishful thinking, but they impressed those individuals who were beginning to work with manufacturers on the appearance of mass-produced items. More important at the time, his comments spoke to New York architects engaged in designing skyscrapers in a modernistic idiom of French decorative elements transformed by the new machine aesthetic.

A few years after the Paris exposition Paul T. Frankl noted angrily that if America could have exhibited a skyscraper, "it would have been a more vital contribution in the field of modern art than all the things done in Europe added together." His own skyscraper furniture paid homage to the unique achievement of American architects, and he concisely stated a philosophy that guided them in adapting the modernistic style. "Simplicity is the keynote of modernism," he noted, but other characteristics included "*continuity of line*, (as we find in the stream-line body of a car or in the long unbroken lines in fashions); *contrasts in colours* and sharp *contrasts in light and shadow* created through definite and angular mouldings and by broken planes." With jazz in mind he concluded that these elements made "a definite rhythm such as we find in modern dancing and music."[47]

Simplicity is relative, but skyscrapers of the late twenties seemed simple compared with the neo-Gothic and neoclassical extravaganzas of a few years before. Extreme height provided continuity of line, enlivened rather than broken up by the common setback motif as a building rose. Unlike Internationalists, who relied on bands of windows for horizontal relief, modernistic architects often used bands of brick or terra-cotta in contrasting colors, which yielded a smooth, machined look. Ornament of the exposition

type generally appeared at the base of a modernistic skyscraper, and modifications occurred as architects became inspired by the machine. Stylized organic motifs prevalent in Paris disappeared, replaced by similarly stylized lightning bolts (electricity), gears, and radio waves, or abstract constructions whose impersonality, precision, regularity, and metallic look suggested by implication the machine (figure 11). Interiors of lobbies revealed a profusion of ornament characteristic of the Paris exposition, but again machine motifs predominated. Historian David Gebhard has noted "the sharp linear angularity" of modernistic ornament in America, concluding that "all of it conveyed the feeling that if it were not produced by the machine it should have been."[48]

Today these skyscrapers seem Gothic with their crenellated towers and their urge to pierce the sky, but to contemporaries they seemed stripped of decoration, almost inhuman in their regularity. Even Hugh Ferriss, a renderer who popularized the style, felt constrained to ask rhetorically if "these masses of steel and glass" did not embody "some blind and mechanical force that has imposed itself, as though from without, on a helpless humanity."[49] Viewed from a distance, the Manhattan skyline suggested the metropolis as a vast social machine, in which human beings merely lubricated its operation. But Ferriss rejected this attitude. His influential volume of renderings, *The Metropolis of Tomorrow*, hinted at limitless potential to be unfolded through the modern spirit and its style. But Ferriss expressed his optimism in a personal style unlikely to awaken similar emotion in others. His monolithic towers of black charcoal, standing alone among misty surroundings devoid of human life, forebodingly insisted on architecture as metaphysics. Contemporaries marveled at his anthill visions of towering structures with multilevel, multilane highways tunneling through them, elevated pedestrian arcades, rooftop airports, suspen-

11. Bronze radiator grill, Chanin building, New York, 1929. Marymount Manhattan College.

sion bridges with skyscrapers for pylons and apartments hanging from cables, and groups of buildings fully sheathed in steel or colored glass (figure 12). The Science Zone of his projected metropolis, with icy reflecting pools surrounding buildings either patterned on machine shapes or resembling phallic quartz crystals (figure 13), epitomized Ferriss's modernistic vision. In describing one of its towers, he waxed portentously poetic: "Buildings like crystals. / Walls of translucent glass. / Sheer glass blocks sheathing a steel grill. / No Gothic branch: no Acanthus leaf: no recollection of the plant world. / A mineral kingdom. / Gleaming stalagmites. / Forms as cold as ice. / Mathematics. / Night in the Science Zone."[50]

In 1930 industrial designer John Vassos and his wife Ruth paid Ferriss the compliment of parody with their *Ultimo*, an illustrated tale of increasing entropy and the freezing-up of the earth. As an ice age engulfed civilization, "buildings dripped with gigantic icicles, awesome and beautiful in the cold rays of the sun." Finally, with a nod to Ferriss, they imagined "the tall and beautiful cities . . . abandoned and desolate, towers of frozen immobility," while the remnants of humanity escaped underground to create an environment whose "infallibility" and "monotony" oppressed its inhabitants, driving some of them to risk the "uncertainties" of interstellar flight. Following Ferriss, John Vassos rendered human figures so indistinctly as to suggest ciphers rather than living individuals. An ultimate machine-age world held no place for people.[51] Not many observers found inhuman or totalitarian implications in the machine aesthetic. Most of its expressions innocently, even whimsically, celebrated such technological innovations as the airplane, the automobile, radio, and

12. Hugh Ferriss. Visualization of a future city, 1929. The Metropolis of Tomorrow.

13. Hugh Ferriss. The Science Zone, 1929. The Metropolis of Tomorrow.

electrical power, all of which provided people greater control over their lives than ever before. If, as Gebhard suggests, designers and architects became obsessed with transportation machines while developing an American version of the modernistic style, then perhaps the Chrysler building (figure 14), completed in 1930, marked the culmination of the style's architectural, pre-industrial-design stage.

The building was originally designed in 1927 by William Van Alen as a conventional setback structure topped by a glass dome,

but the plans were changed when they were acquired with the site by Walter P. Chrysler. A thirtieth-floor brickwork frieze at the base of the main shaft depicted stylized automobile wheels, hubcaps, and fenders, punctuated at the corners by sleek eagle gargoyles resembling huge radiator caps. Its crowning glory was its four-sided cap of receding semicircular arches, faced with platinum-colored steel, each arch being broken up by spiky triangular windows. Topped with a gleaming communications mast, the building resembled a slender, silvery rocket and in-

*14. William Van Alen. Chrysler building, New
York, 1930. Cervin Robinson.*

spired illustrators of science-fiction pulp magazines as they depicted cities of the future. In the lobby, murals portrayed airplanes in flight over a world map, dirigibles, speeding automobiles, and the shining Chrysler building itself, as if it were a capstone to modern civilization. Gaudy and perhaps vulgar, the Chrysler building expressed a cheerful side of the machine aesthetic far removed from Ferriss's ponderous meditations, although he illustrated it in his *Metropolis of Tomorrow*. When he made his rendering, he included the fragile framework of its uncompleted crown, around which shone a halo of light, a penumbra of the sun rising behind—symbolizing the dawn of a new age.[52]

Faith in a future of technological progress also appeared in *The New World Architecture* by Sheldon Cheney, a drama critic branching out into architecture and design. Humanity, he thought, was emerging from the initial stage of the industrial revolution, marked by slavery to machines and a rape of the environment. Now that mankind was gaining control of the machine, even the masses could reap the results of their ancestors' servitude: "cleanliness, education, travel, and enjoyment of the arts." The machine would be made "so easy, so efficient, so noiseless, that we rise beyond it to enjoy those serenities, those spiritual contacts, those pleasures of quietness, that enriched life (for a few) before the machine era." With others of his time Cheney thought architecture should express the dominant theme of the age—the machine. But he found "modernistic design . . . too easy, too shallow, too soft" because it relied on surface ornament. Cheney admired the Weissenhof suburb but considered it a negative statement. He found its proponents too doctrinaire to permit "individualism of expression." Rejecting Le Corbusier's vision of human beings as interchangeable thinking machines, he declared that "the scientific regimentation of living must be *for the liberation of man's individuality, of his cre-*

ativeness." For Cheney the exemplar of the machine aesthetic was Eric Mendelsohn, "the architect who most rationally and most intensively—in projects and pictures, at least—made buildings expressive of the 'feel' of the machine," combining "expressiveness of the intended use of the building" with "an expressiveness of materials."[53]

Americans knew little of Mendelsohn until 1929, when the Art Center's Contempora Exposition of Art and Industry presented his visionary sketches, photographs of completed works, and a few models, incongruously displayed with typical interior ensembles by other designers.[54] In a radio talk during the exposition Frank Lloyd Wright declared Mendelsohn's work a "romantic" and "powerful realization of the picturesqueness of our special machine brutalities."[55] Aldous Huxley also expressed admiration for Mendelsohn. Attacking the antiseptic "aesthetic puritanism" of Le Corbusier, he praised Mendelsohn's "architectural romanticism," achieved by "strictly contemporary technique and contemporary artistic formulae."[56]

As early as 1914 Mendelsohn had begun to formulate ideas about a new architecture. He wrote his fiancée that steel and reinforced concrete would be the construction materials of "the new formal expression . . . the new style."[57] Joining the German army a year later, he found his architectural vision stimulated by gun emplacements, observation posts, and shelters of reinforced concrete— many with low horizontal lines and rounded corners (for deflecting shells) similar to those in his "trench sketches" of 1917. Before he had executed a single commission, he created the sketches that later caused a stir at the Contempora Exposition.[58] Depicting massive industrial and commercial buildings—a harbor grain depot, a film studio, a heavy equipment plant, an optical factory (figure 15), and so on—they relied on sweeping arcades of semicircular arches, curving bands of windows, and a plethora of S-curves to produce an effect of motion.

Mendelsohn partially realized these visions in his Einstein Tower at Potsdam (1920–1924). More influential because more practical were his commercial buildings of the twenties. Such buildings as the Schocken store of Stuttgart (1926–1927), the Petersdorff store of Breslau (1926–1927), and the Universum theater of Berlin (1927–1928) provided movement in their flowing horizontal curves (see figure 16). Although Mendelsohn, unlike American architects and industrial designers who followed him in the thirties, did not explicitly reproduce airplane and automobile forms in his streamlined designs, he did succeed in expressing the essence of the transportation machine—speed.

15. Eric Mendelsohn. Sketch for an optical factory, 1917. Wolf Von Eckardt.

With Le Corbusier and American modernists he found "the machine" to be "the point of departure for the new culture." Mechanization, he thought, would affect every "manifestation" of emotion and thought, bringing architecture in line with the machine as designers sought "to achieve complete unity between functional form and artistic form."[59] Only someone with "no rhythm in his body" could fail "to understand the metallic rhythm of the machine, the humming of the propeller, the huge vitality which thrills us, overjoys us and makes us creative."[60] His designs took into account "the influence of speed which dominates our lives." Mendelsohn noted that "the man in a motor-car, or in an aeroplane, with his tense vision, has no longer any leisure to notice details and petty decoration" but "requires great masses and bold lines."[61] From design intended to accommodate the mentality of a machine operator it was not far to design based on machines themselves.

As American designers of the early thirties searched for a style expressive of the new tempo of life, they turned naturally to Mendelsohn's example of a more exuberant machine aesthetic than Le Corbusier's. Unfortunately for the critical reception of streamlining, they could apply Mendelsohn's vocabulary of curves as superficially as

16. Eric Mendelsohn. Schocken department store, Stuttgart, 1926–1927. Wolf Von Eckardt.

French-inspired angular decorative motifs, but even the slickest streamlined design witnessed a designer's belief that the public desired visual confirmation of technological progress. In the late twenties, however, the jagged modernistic style remained the major American expression of a machine aesthetic, while Mendelsohn's visions remained just that. Owing to the conservatism of the American design establishment, the first industrial designers turned to architects for inspiration. While New York modernists like Ferriss and Hood offered machine motifs that could be applied to product design, Le Corbusier provided an uplifting vision of society rationally harmonized and functionally organized in the machine's image.

In moving from architecture to product design, the machine aesthetic assumed new dimensions. Architecture in general remained the conceptualizing end of the construction industry, a bastion of craftsmanship and one-of-a-kind production. Industrial design, on the other hand, concerned itself with mass production. Designers could not escape the fact that they created *for* the machine, providing patterns to be mechanically reproduced endlessly (until the next model change). A machine aesthetic thus involved concern for the machine as a production tool as well as a source of inspiration. Industrial design involved a social responsibility more insistent than that of the architect, who had to satisfy only a single client. As Frederick Kiesler declared, "THE NEW ART IS FOR THE MASSES," and he expected to popularize it through the medium of the store.[62] Designers bore the burden of ensuring that products honestly expressed the modernity implied in the machine aesthetic. Bach pointed out that machine products were only as good as their designers. Always an evangelist, he viewed "the machine as an economic factor, design as a spiritual necessity, the one to produce the other and make it available in quantity to the mass."[63] The artists and designers who deserted prior careers in the late twenties to answer business's anguished call attempted in different ways and with varying success to follow Bach's injunction.

The New Industrial Designers

It is as absurd to condemn an artist of today for applying his ability to industry as it is to condemn Phidias, Giotto or Michelangelo for applying theirs toward religion.
—Norman Bel Geddes

Industrial design was born of a lucky conjunction of a saturated market, which forced manufacturers to distinguish their products from others, and a new machine style, which provided motifs easily applied by designers and recognized by a sensitized public as "modern." Without an economic impetus, modernistic design would have remained a preoccupation of the elite. And without the aesthetic provocation of the Paris exposition, designers would have lacked a common focus for their appeal to the public. Dictated as before by eclectic whim, product designs would have lacked the coherence of a common style. The public indeed desired novelty, but a coherent machine style provided as well a sense of security amidst rapid change, a feeling that everything was under control. Without consciously noticing the resemblance of a stepped radio cabinet to a setback skyscraper or of an electric fan to an airplane motor, a consumer received reinforcement for purchases from subliminal perceptions of continuity in the man-made environment. Without this web of cross-references provided by a common style, industrial design would not have demonstrated its usefulness to businessmen.

Not only the public yielded to this vision of harmony and control. Designers themselves felt the attractions of images they had created. Ambivalent about loyalties to art and business, they sometimes rejected their commercial role, selling products, and instead saw themselves as pioneers remaking society by bringing its artifacts in line with modern times. Tension between commercialism and social service ran through their published remarks and revealed itself in compromises necessary for success. Many industrial designers found it necessary to justify their apparent defection to mammon's camp, but they did so in the beginning as isolated individuals. None of them knew, as business needs and modernistic design began to combine synergistically, that they would become part of a movement. In search of employment, they simply drifted into a new profession, whose purpose was the application of art to industry. Eventually about fifteen designers emerged as prominent members of the profession. Only four or five developed methods efficient enough and staffs large enough to accept commissions from major corporations. But in the early years a variety of approaches rendered the profession fluid.

Both Fish and Fowl

A manufacturer often made initial contact with a commercial artist or decorative craftsman by obtaining a name from an advertising agency, the Art Center, or the Metropolitan Museum. At first the typical manufacturer venturing into art in industry wanted his product superficially "styled" and requested a few sketches that would guide production engineers in "beautifying" a product whose functional specifications were set. Such sketches often proved impractical and provoked hostility because the typical novice industrial designer knew nothing about industrial materials and processes. But at least one succeeded with this approach. Lurelle Guild, who graduated in fine arts from Syracuse University in 1920 and spent several years illustrating interiors for women's magazines, allotted only a few hours for each job. After sketching a dozen ideas, he sent his client the four best for consideration. Guild boasted that he redesigned a thousand products a year; but those designers who emerged as heads of large design offices attributed success to their refusal to provide superficial sketches and to their concern for function as well as appearance.[1]

As design of appearance became more important in selling products, some larger firms invigorated their art departments by hiring designers familiar with modernistic style. In 1928 General Electric hired Ray Patten to supervise in-house design of domestic appliances, while Donald R. Dohner, who had taught design at Carnegie Institute of Technology, took the same position at Westinghouse in 1930.[2] George Sakier, a Columbia-trained engineer who had studied painting in Paris and had worked as commercial illustrator, art director for *Harper's Bazaar*, and interior decorator, became director of a Bureau of Design in the American Radiator and Standard Sanitary Corporation. Concentrating at first on luxury plumbing fixtures, he turned in 1929 to styling inexpensive lines of products.[3] Despite these examples, most designers worked as independent consultants.

Many drifted into design as an extension of previous occupations. Often they could not free themselves from earlier work habits. John Vassos, a Greek who came to America in 1919 and studied at the Art Students League, did not become a major designer because he worked alone as he had done as an illustrator rather than with a staff of draftsmen and engineers.[4] Vassos was typical of designers who began as commercial artists and found themselves asked to design labels, packages, and an occasional small appliance whose housing functioned as a package. Joseph Sinel, a New Zealander who arrived in America with a knowledge of calligraphy and typography gained in Britain, moved from advertising to a career in industrial design that centered on packaging (see figure 17).[5] George Switzer, an Indiana native educated at the University of Illinois, worked as art director of an advertising agency until resigning in 1929 to open an industrial design office. Most of his work consisted of packaging, including creation of a standard image for Hormel meat products and the well-known Wonder Bread wrapper.[6] Gustav Jensen, who arrived from Denmark in 1918, gained publicity with a Monel metal sink designed for International Nickel in 1931 (figure 18), but he worked primarily on packaging, cosmetic bottles, magazine layouts, and commercial brochures.[7]

The most influential packager was Egmont Arens, who publicized industrial design with

*17. Joseph Sinel. Packaging, 1930. Crain
Communications, Inc.*

numerous articles in trade journals. Born in
Cleveland and educated at the universities of
Chicago and New Mexico, Arens began his
career as a sports editor in Albuquerque and
moved to New York to edit such magazines
as *Creative Arts* and the original *Playboy*. A
few months before the stock market crash he
joined the art department of Calkins and
Holden. There he started the Industrial Styling Division, which provided packaging
design (see figure 19) and an occasional
product design for agency clients.[8]

Although these men called themselves industrial designers and figured in publicity for
the new profession, their emphasis on packaging, which required little support staff,
precluded emergence as major product designers. But they aided the movement by
introducing modernistic motifs in the most
ubiquitous commercial form, the package.

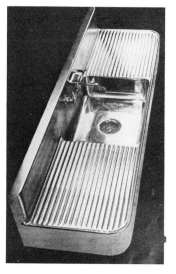

*18. Gustav Jensen. Monel metal sink for
International Nickel, 1931.* Pencil Points.

19. *Egmont Arens. Crayonex packaging, c. 1930. Arens Collection.*

Their work, at one end of a scale running from packaging through consumer products to commercial architecture, revealed industrial design's fundamental relationship with advertising. Since few functional considerations affected their work, they concentrated on selling goods by giving their packages superficial images of modernity. The packagers' success in boosting sales speeded manufacturers' acceptance of product redesign itself. Since a new box or label required no retooling, executives readily accepted advertising agencies' demands for package redesign but often balked at redesigning products. With the efficacy of new packaging demonstrated, manufacturers accepted the idea of redesigning a product to provide its own advertisement.

A second group of industrial designers, whose backgrounds lay in decorative craftsmanship, also found horizons limited by early experience. Peter Müller-Munk arrived from Berlin in 1926 and worked as a silversmith at Tiffany's. Gradually he shifted to such mass production jobs as a line of dishes for a New York department store, but his work remained solidly in the decorative tradition throughout the thirties.[9] Another designer, Gilbert Rohde, started out as an advertising illustrator specializing in interiors.

After visiting Europe in 1926 to study modern interiors, he began making tubular tables for the custom market and became head designer for the Herman Miller furniture company. Although Rohde designed appliances for General Electric in the thirties, he remained a furniture designer.[10]

Another of this group, Russel Wright, was born in Lebanon, Ohio, and studied at the Cincinnati Art Academy and the Art Students League before becoming a stage designer. His entry into decorative crafts came around 1927 when the owner of a New York custom furnishings shop suggested that papier-mâché animals he had made as stage props would sell if cast in metal as bookends. Soon his own workshop was producing wooden bowls, cutting boards, and kitchen utensils of spun aluminum. Although from 1929 he designed pianos and radios for the Wurlitzer Company, he concentrated on furniture and ceramics, at first for the custom trade and later for mass production. In 1932 his letterhead identified him as engaged in "Modern Accessories—Styling," but his name figured in articles devoted to industrial design. As Wright recalled in 1962, he worked intuitively with no standard method, which suggests why he never became a major designer in industry.[11]

Although these designers helped Americanize the modernistic style and were considered leaders of the industrial design movement, their roots in the decorative art tradition made it difficult to design products for mass production and distribution. Like designers who specialized in packaging, they tended to work alone or with an assistant. They never developed organizations large enough to command the patronage of big business. They might have argued that they were artists, not businessmen. In any case, they did not contribute in a major way to forming and rationalizing the field of industrial design so that manufacturers could grasp its uses.

The four designers who did emerge in the thirties as acknowledged leaders and who regularly won commissions from large corporations had no ties to the traditional decorative crafts. Walter Dorwin Teague had worked for more than fifteen years as an advertising illustrator, Norman Bel Geddes and Henry Dreyfuss had made successful careers as stage designers, and Raymond Loewy had become one of New York's leading fashion illustrators. They too answered industry's call without realizing they were shifting careers. But for a variety of reasons they evolved similar methods of operation and developed the first full-fledged industrial design offices. They inspired others who decided to enter the profession on their own and also provided experience for young draftsmen, designers, and architects who institutionalized industrial design in the forties and fifties by opening their own offices. The lives and work of these "big four" New York designers provide an index of the origins, purposes, and early accomplishments of industrial design.

A Typical Present-Day Businessman

When the Cheneys reviewed Walter Dorwin Teague's work in their enthusiastic *Art and the Machine*, they praised him as a typical businessman "whose business is design." Teague (figure 20) was a "realist," concentrating on "the organization of a manufactured product to increase its desirability, and hence its sales." Prospective clients received his "Outline of Industrial Design," listing his goals as improved appearance, greater ease of service, and economy of manufacture. Every design job included a study of a product's functional and structural elements as well as its manufacturer's production capabilities, sales methods, and advertising techniques.[12] A man of tact and diplomacy, Teague delegated design work to subordinates and focused energy on cultivating clients and negotiating with them. From the start Teague's businesslike manner impressed manufacturers who feared the imag-

20. *Walter Dorwin Teague and his son, W. Dorwin Teague, Jr., c. 1939. W. Dorwin Teague, Jr.*

ined vagaries of long-haired artists. He received his first commission—for two cameras—late in 1927 from Adolph Stuber of Eastman Kodak, who had obtained his name from Bach of the Metropolitan Museum. Teague balked at supplying mere sketches and insisted on inspecting the factory and on collaborating with Kodak's production engineers to ensure feasibility. This pragmatic approach to design won him Kodak's business for the next thirty years (see figure 21) and quickly brought him a success denied to designers unable or unwilling to address business needs and methods.[13]

Nothing in Teague's origins suggested a career in design or in business. Born in 1883, the son of a circuit-riding Methodist minister, Teague grew up in the small town of Pendleton, Indiana. Shortly after graduating from high school in 1902, with seventy dollars in his pocket, he left home for New York. There he supported five years of night classes at the Art Students League by painting signs and drawing for mail-order catalogues. Aspiring to follow in the footsteps of Maxfield Parrish and Howard Pyle, he produced a few magazine illustrations and book designs, but his career faltered until 1908, when he joined the art department of Calkins and

21. *Teague. Kodak Baby Brownie camera, c. 1935.*
Pencil Points.

Holden. Years later Teague told Calkins that he "considered it one of the most fortunate happenings of my life . . . that I came under the influence of you and Ralph Holden at the very beginning of my professional career."[14] There is no reason to doubt his sincerity.

From his employer Teague learned of the commercial artist's duty to improve public taste and thus help refine the national culture. As a novice advertising artist, however, he sometimes doubted the propriety of selling out to business instead of following art for art's sake. This concern surfaced in a paper that Teague read to colleagues around 1910. He began defensively by exclaiming, "Let me raise the voice of one of the 'many incompetents' who 'contrive to make a living.'" Aware of losing status by entering commercial art, he mentioned that he continued to create paintings, etchings, and woodcuts for friends. Because meeting business demands often required lowering personal aesthetic standards, he continued, "what is had in our product is not always our fault but we do inward penance for it." Dis-

satisfied, he and his fellows strove "in devious ways to force a higher standard of artistic work upon a none too cordial market" and looked for "the dawn of a tomorrow when the public will not only require that all its art shall be worthy of the name, but when there shall be *Artists* worthy too."[15] His crusade for commercial art of quality continued as he later entered industrial design.

Around 1910 Teague began an intensive study of seventeenth- and eighteenth-century French history and culture. His acquired "sympathy for classical forms"— as a journalist phrased it—expressed the precision of his restrained personality, and even during the modernistic twenties he remained in temperament "still a classicist and traditionalist."[16] After opening shop as a free-lance artist in 1912, Teague began using classical motifs in his advertising work. His "Teague borders," decorative frames enclosing illustration and copy, provided ads for luxury goods with a refined dignity. A typical example, created for Pierce-Arrow around 1915, ironically resembled an eighteenth-century memorial, its classical pedestal supporting not a bust or funeral urn but a painting of a Pierce-Arrow car enclosed within a convex bay-leaf molding.[17] His work remained unchanged until 1920, when his borders reflected a new design theory supposedly derived from the Greeks and compatible with his classicism. This theory, Jay Hambidge's "dynamic symmetry," not only transformed Teague's borders but also influenced his later conception of industrial design.

Hambidge's ideas probably reached Teague through articles published in *The Diagonal* in 1919 and 1920 and reprinted in 1926 as *The Elements of Dynamic Symmetry*. A former reporter and magazine illustrator, Hambidge considered contemporary artists "on the threshold of a design awakening" that would replace "pilfering" of historical styles with "a healthy and natural expression of the aspirations of our own age." Unlike some contemporaries, however, he deplored "the modern tendency to regard design as purely instinctive." Asserting intelligence over intuition, Hambidge offered a design science derived from study of the classical Greeks, who "had a clear understanding of law and order and a passion for its enforcement." Working from measurements of the Parthenon, Hambidge derived a geometric method for constructing proportional relations between different parts of a design in order to maintain a "just balance of variety in unity." According to him, dynamic symmetry reflected knowledge of eternal laws of proportion embodied in natural phenomena—such as the arrangement of leaves on a stem and the spiral of a shell. First discovered by the Egyptians, dynamic symmetry was rationalized by the Greeks but then lost until Hambidge's rediscovery. In essence, Hambidge called for a new golden age, based on the principles of the old, when human constructions would again harmonize with one another and with the eternal laws of nature.[18]

Hambidge's classicism appealed to Teague, who abandoned eighteenth-century motifs for a logarithmic spiral derived from the golden mean of dynamic symmetry. The issue of *Vogue* for August 1, 1922, contained striking examples of pre- and post-Hambidge borders. An ad for Fatima cigarettes (figure 22) contained his usual rectangular frames, moldings, garlands, an urn, and other classical motifs, while an ad for Phoenix Hosiery on the facing page (figure 23) bore a mathematically precise tracery of vines, leaves, and tendrils, entwining through a multitude of curlicues inspired by Hambidge's spirals. In the next few years Teague explored permutations of organization and layout of this ornamental motif.

More important, he also adopted the general tone of Hambidge's design philosophy as his own. The title of a book he published in 1940—*Design This Day: The Technique of Order in the Machine Age*— revealed an orientation similar to Hambidge's. One chapter, "Rhythm of Proportion," expounded dynamic symmetry without modification. According to Teague, Hambidge's treatise yielded "certain relatively simple and highly useful tools" with which "we may adjust the internal relationships of our designs with some of the accuracy the Greeks attained." By using the golden mean throughout the parts of even a consumer product, a designer could "permeate every detail of the structure" with a common rhythm, yielding "a serenity and rightness and completeness in the composition." And, as Hambidge had also asserted, "by integrating our own work rhythmically we are repeating the structural scheme of the universe in which we live."[19] Eventually, Teague conceived of the industrial designer's sphere of influence as the entire man-made environment. In *Design This Day* he echoed Hambidge's respect for classical Greece, a civilization that resonated with common harmonies in arts, sciences, and social arrangements, and he stressed the necessity of designing and engineering such a society for survival of the human race. He looked

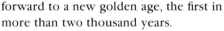

22. *Teague. Advertisement for Fatima cigarettes, 1922*. Vogue.

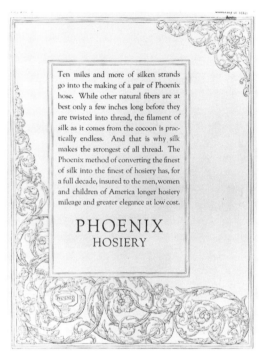

23. *Teague. Advertisement for Phoenix Hosiery, 1922*. Vogue.

forward to a new golden age, the first in more than two thousand years.

Teague began thinking along these lines in the mid-twenties as a result of his classical studies of French civilization and dynamic symmetry. He grew bored with repetitive advertising borders and looked for an expansive design role. In a speech to the Art Directors Club in 1925 he advised colleagues to study "the close relationship between all the arts of design in any given period and see how buildings, furniture, clothing, textiles, even common utensils of everyday life are all expressive of the same taste and the same fashion."[20] Teague found this unity of style in historical Greece and France but considered it as yet only a prescription for his own society. Attracted by developments in architecture in Europe, he traveled there in 1926 to "burn his mental bridges behind him."[21] Inspecting the work of Le Corbusier, Walter Gropius, Robert Mallet-Stevens, and the modernistic decora-

tors of Paris, he returned home "a wholehearted disciple of 'modern' design."[22] The various examples of aesthetic ferment in Europe coalesced in Teague's mind to produce an impression of a single modern style, capable of leading to a civilization marked by harmony in all its parts.

Soon after arriving home, he replaced the French period decor of his office with a restrained mixture of modern elements—flat white walls, tubular chairs, and cabinets in the Paris style.[23] Entering into the new tempo with customary diligence, he compiled massive files of clippings from two French design magazines, *Mobilier et Décoration* and *Art et Industrie*.[24] In mastering the new idiom, Teague demonstrated a scholarly approach that had yielded his advertising borders. Never an originator, he exhibited solid knowledge of what others had created, but his recombinations continued to reflect classical restraint, decorum, and understatement. As late as 1931 his de-

sign for an Eastman Kodak shop at 745 Fifth
Avenue (figure 24) revealed indebtedness to
the exposition style, but he maintained
simple geometric themes. Flat stepped-out
pilasters and stylized chromium capitals
expressed his style in those transitional
years: modernism shaped by a classical
vision. This restrained vision won the confidence of businessmen and, on the eve of
the Depression, was rewarded with a commission to design the Marmon 16 luxury
auto (figure 25), which marked his emergence as a full-fledged industrial designer.

If Teague became the most successful
industrial designer, measured by volume of
completed work, he did so because he
adopted business procedures and adapted
himself to his clients' professed need—increased profit—without greatly compromising aesthetic integrity. Raymond Loewy once
complained that Teague never discussed
anything besides acquiring more clients,
making more money, and pumping Loewy
for information about his own clients.[25] But
these concerns seem usual when two rivals
talk shop. Teague never lost sight of his
mentor, Earnest Elmo Calkins, from whom
he gained a faith that advertising—and by
extension industrial design—could help
build a harmonious civilization. Working
through business for social betterment provided Teague with sincere motivation for his
work. A publicist in 1934 described this concern as "social, not individual," because
Teague believed "that American advertising
as a corollary of mass production will raise
the level of human comfort."[26]

Teague maintained his conservative faith
in the social uses of business in general and
design in particular through the Depression,
voted Republican, and expressed his beliefs
in a letter written in 1940 to solicit a vote
for Wendell Willkie. With his clients Teague
believed in "the great future of America
under a system of free enterprise" and
thought "that our greatest national problem
is the raising of the standard of living for all
the people." Businessmen would be "in-

*24. Teague. Interior of Eastman Kodak camera
shop, New York, 1931.* Architectural Record.

25. Teague. Marmon 16 automobile, 1932. Pencil
Points.

finitely better off" if everyone "could be happy and prosperous." Ten years of Depression and Democratic rule had stalled the nation on "the greatest frontier of all: the frontier beyond which lies a land of diminishing ugliness and disorder where our people can live decently, wholesomely and graciously." Already, he continued, scientists, engineers, and designers had "made forays into this promised land and brought back samples of its fruits." Their "charts and maps" would direct the nation as soon as a new Republican administration had taken the helm.[27] One might conclude that Teague merely echoed his industrialist employers, but his faith in business went back to the resolution of his early struggle over wasting his talent by becoming a commercial artist. To Teague, art and industry proved no contradiction.

26. Norman Bel Geddes (second from right) on a "secret mission" for the Graham-Paige Co., 1928. Geddes Collection.

The P. T. Barnum of Industrial Design

While Teague epitomized the designer as businessman, his colleague Norman Bel Geddes (figure 26) seemed more of a creative genius. His reputation as "bomb-thrower Geddes, the man who has cost American industry a billion dollars" for retooling, provided industrial design with a mystique essential to its development.[28] The breathtaking streamlined visualizations in his book *Horizons* (1932) cornered the market on Sunday-supplement views of "the world of tomorrow." By popularizing streamlining when only a few engineers were considering its functional use, he made possible the design style of the thirties. His career had more immediate professional and cultural impact than those of more practical colleagues.

Like Teague, Geddes, born in 1893, came from a small town, Adrian, Michigan.[29] When he was about thirteen, his father died, leaving behind little money, but his mother held her family together and imbued her son with a sense of purpose, a desire to succeed, and an interest in art and the theater. Geddes

never returned to school after expulsion from the ninth grade, though he did briefly attend the Chicago Art Institute. Beginning in 1913, he worked in several Detroit advertising agencies—illustrating automobiles in a dappled impressionist style. Stage design, originally a hobby of model-making, soon became his main interest. By 1916 Geddes had convinced Aline Barnsdall to let him design the stage of her Los Angeles Little Theatre, for which Frank Lloyd Wright was designing a building. Even then Geddes's ego proved too expansive for collaboration, and the project fell through, but in a few years Geddes had become a leading stage designer. As popularizer of an expressionist "new stagecraft," Geddes believed with its theorists "that the stage setting, as a work of art in itself, should express the dominant mood or emotion of the play."[30] Effective theater required total integration by a single designer of script, set, lighting, color, music, and theater architecture. Such ideas translated well into the terms of industrial design.

From his brief association with Wright, Geddes retained a feeling that stage design

was not far removed from architecture. His
work grew monumental. In 1919 he created
an opera set in which a ship with forty cast
members was launched down a ramp of ball
bearings. For *The Miracle* in 1923 he rebuilt
the interior of New York's Century Theatre
as a medieval cathedral. His most famous
project, a spectacle based on *The Divine
Comedy*, was not produced because it required a special theater.[31] In 1924 Geddes
met Eric Mendelsohn, who gave him a sketch
of his Einstein Tower (figure 27) and a copy
of his *Structures and Sketches*. Geddes
scored heavily a passage in "The Problem of
a New Architecture." From Mendelsohn,
Geddes learned that new architectural forms
would come not from tradition but from engineering developments "in the machine and
in engines of transportation."[32] More important, Mendelsohn's visionary sketches
eventually led Geddes to popularization of
streamlining as an industrial and architectural style. Geddes introduced a catalogue of
Mendelsohn's show at the Contempora
Exposition of Art and Industry in 1929.
Architecture, Geddes wrote, more than any
other art "expresses . . . the spirit of the time
in which it is created." Despite this worldly
connection, "its contemplation relieves the
commonplaceness of everyday activity and
leads toward more ultimate possibilities."
Echoing modernist platitudes, he insisted
that creative architecture "demands freedom
of thought, freedom from dogma, freedom
from history." And finally, after commenting
on architecture's sculptural quality, he concluded on a revealing note by stating that the
ideal architect "with the dramatist's instinct
. . . adds the emotional quality that attracts
and inspires humanity for all future time."
His final sentence suggested Mendelsohn as
this ideal builder, but the theatrical reference revealed his own aspiration to the
laurels he bestowed.[33] Geddes was beginning
to think of himself as an architect. Not until
angry unemployed architects forced cancel-
lation of nine architectural designs for the
Chicago Century of Progress Exposition on

27. *Eric Mendelsohn. Einstein Tower sketch
originally dated 1919 and presented to Geddes by
the architect on November 25, 1924. Geddes
Collection / Mrs. Eric Mendelsohn.*

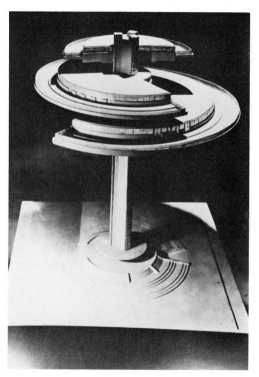

28. *Geddes. Model of proposed revolving
restaurant for Century of Progress Exposition,
1929–1930. Geddes Collection.*

the ground that Geddes lacked certification did he realize that the profession was closed to him (see figure 28).[34]

An aura of optimism surrounded Geddes as he moved toward industrial design. In 1924 he roasted the commercialization of the arts and described his time as the "climax of the iron age, the age of metal—of almighty dollar of steel for every purpose." But he sensed an opposing "art consciousness" building up that would soon erupt to produce an age aesthetically surpassing the Renaissance. He knew from Mendelsohn that the new movement would incorporate elements of an industrial world that had once produced ugliness. Industrial design would eventually prove a dialectical synthesis of two forces that had seemed irreconcilable. A maxim formulated by 1924—"the imagination creates the actual"—revealed his faith in the power of projections of the future to shape the course of civilization.[35]

Geddes's expansion beyond stage design pioneered a generalist principle of industrial design: the application of ideas, materials, or processes from one field to another. A walk down Fifth Avenue in 1927 convinced him that he could provide drab shop windows with "come on" by treating a window "as a stage, the merchandise as the players, and the public as the audience."[36] His expressionist window for Franklin Simon (figure 29), the beginning of a two-year association, focused on projecting a storewide image of sleek, refined elegance rather than on selling specific merchandise. A later display of luggage (figure 30) revealed Geddes's interest in fast transportation machines in his use of

29. Geddes. First display window for Franklin Simon, 1927. Geddes Collection.

30. Geddes. Franklin Simon display window, c. 1927. Geddes Collection.

elements from A. M. Cassandre's famous "Etoile du Nord" railway poster (figure 31). Elegance became identified with the machine-age tempo.

In 1928 Geddes hired Frances Resor Waite, a young woman who admired his theater work, to oversee the Franklin Simon windows. She became his second wife in 1933 but more immediately helped him professionally by introducing him to her uncle, Stanley Resor, president of the J. Walter Thompson advertising agency. Soon after meeting Resor, Geddes began pursuing a "wholesale designing idea."[37] In addition to providing him with clients, the agency commissioned a combined auditorium and conference room for its office in the Graybar building (figure 32). Geddes embraced "the new simplicity" to create a room "machine-like in its efficiency."[38] Although its deep

31. A. M. Cassandre. Railway poster. Klaus-Jürgen Sembach.

32. Geddes. Auditorium for J. Walter Thompson, New York, 1929. Geddes Collection.

33. Geddes. Metal bedroom furniture for Simmons Co., 1929. Geddes Collection.

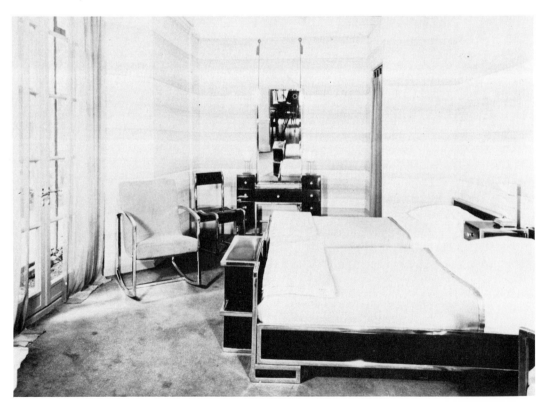

horseshoe chairs reflected the exposition style, he avoided exotic ornament for severe accent strips of brass and black glass on off-white walls. Gray carpet was relieved only by turquoise upholstery and drapes. With concealed lighting operated by twenty-second dimmer switches, omnipresent phone jacks, an automated movie screen, and a ventilation system replacing contact with the outside environment—all of which eliminated random irritations understood then as "friction"—the room embodied control. It expressed the function of an agency—lubricating the consumer society.

None of Geddes's pre-Depression designs proved commercially successful, but they won him flamboyant publicity. A metal bedroom suite for the Simmons Company, a J. Walter Thompson client (figure 33), won praise among the fashionable for its modernistic black lacquer and chrome, but the firm soon returned to a more popular imitation-wood finish for mass appeal.[39] For another agency client, the Toledo Scale Company, Geddes's office worked on a grocery-store counter scale, but it never became more concrete than Al Leidenfrost's luminous rendering (figure 34). Glowing with an aura of reflected light along its sculptural lines, the scale in his illustration was a machine-age icon and proved more effective as publicity than would a photograph of an actual product. Before the company became disillusioned with Geddes, however, it commissioned him to design an entire factory complex, for which it had scrapped plans already completed by architect Harvey Wiley Corbett.[40] Ordered to create something "entirely different, even weird looking,"[41] Geddes opted for a pastiche of Wright's prairie architecture and Mendelsohn's streamlining (figures 35 and 36). Geddes later referred to the project—which would have included an airport for executives and athletic fields for employees—as "one of the most satisfying experiences of my life."[42] But it too produced only a sheaf of impressive renderings and a reputation for

34. Geddes. Proposed grocery counter scale for Toledo Scale Co., rendering by Al Leidenfrost, 1929. Geddes Collection.

Geddes as the man who, when asked to redesign a product, went on to design its factory.[43]

A final pre-Depression project, equally abortive, foreshadowed Geddes's role in the thirties as a visionary of the future. Attempting to rationalize the annual model change with a plan possibly inspired by Creange's three-phase system, automaker Ray Graham of the Graham-Paige Company commissioned six designs, two for cars "which can be sold immediately," two for cars "with mildly radical lines," and two for cars "with lines five years ahead of those used in cars today."[44] The plan was soon modified, calling for five successively more radical designs to be introduced over a span of five years in order to accustom the public to change. At this point Geddes was "not interested in designing a car that looks like a cigar or an aeroplane" but concentrated on designing "a single unit of uninterrupted, flowing lines."[45] Always a dreamer, he conceived of the fifth in the series as the "Ultimate Car," technically and aesthetically perfect (figure 37).[46] By the time Geddes discussed the

35. *Geddes. Rendering of administration building in factory complex proposed for Toledo Scale Co., 1929. Geddes Collection.*

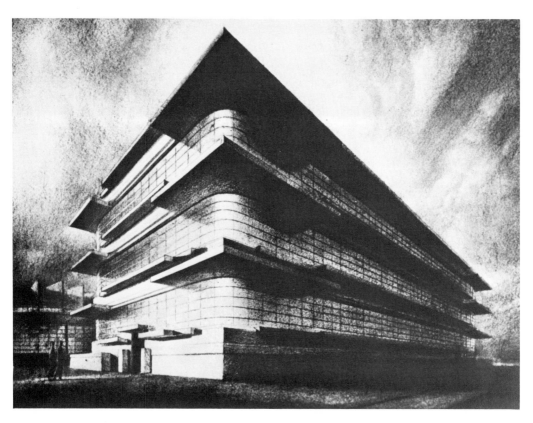

36. *Geddes. Rendering of assembly building in factory complex proposed for Toledo Scale Co., 1929. Geddes Collection.*

37. Geddes. The "Ultimate Car"–most radical in a series of five unrealized designs for Graham-Paige Co., 1929. Geddes Collection.

design in *Horizons*, his own streamlining had so far surpassed it that he misrepresented the Ultimate Car as the least radical in the series of five.[47] None were produced, the stock market crash having ended such risky experiments.

A natural showman, Geddes learned quickly how to gain publicity even from his failures and thus boost his own reputation as well as that of his fledgling profession. Renderings and photographs of models published in trade journals and in his own *Horizons* provided manufacturers with dramatic evidence of what could be done through redesign, even though in many cases it had *not* been done. In his method of selling to businessmen as in his way of selling to the public, image often took precedence over reality. By the beginning of the Depression Geddes had switched most of his energy from theater to industrial design. Through "a quite natural evolution," as he described it, he had entered the field of "utilitarian art" and dedicated himself to turning "frankly commercial objects" into "satisfying objects of beauty." Defending his new career in a drama magazine, he wrote that "an artist never strays from his natural path so long as

he is earnestly attempting to create beauty, whether it is in a painting, a poem, a setting for a drama, the building of a factory, or the manufacture of a chair."[48] His eventual success as an industrial designer depended precisely on his willingness—desire, in fact —to enter fields about which he knew nothing. Geddes himself found valuable "an increasing lack on my part of knowing about anything." A specialist's expertise, he thought, rarely matched the creativity of a generalist who based quick decisions on intuition.[49] If this spontaneity often yielded impractical results, it also put him on the cutting edge of his profession. A short, pugnacious individual, Geddes confessed himself a "complete failure" at relaxing and traveled far on sheer exuberance.[50]

A Good Human Being

In his autobiography Buckminster Fuller stigmatized industrial design as "the greatest betrayal of mass communication integrity in our era." Some years later, however, he praised Henry Dreyfuss (figure 38) as a leader in a "design revolution" capable of bring-

38. Henry Dreyfuss, 1934. Design.

ing the human race to "an utterly new omni-successful relationship to universe."[51] For Dreyfuss he made an exception. Although Dreyfuss studied under Geddes, his conception of industrial design resembled Teague's. When asked for speculations on the "future appearance of a few familiar products," Dreyfuss refused because such "poppycock publicity" harmed the profession.[52] He gained clients by recognizing "the value of understatement."[53] In the early thirties John S. Tomajan of the Washburn Company visited several designers for help with a line of kitchen utensils. Most designers offended him with "an air of cocksureness," but Dreyfuss's "natural, self-effacing manner" won him over.[54] When Dreyfuss did play the showman, as with an automated drawer that swung from his desk to offer cigarettes, he stuck to small effects.[55] Cultivating the image of a businessman, he made a trade-mark of his plain brown suits, worn to counteract the stereotype of the artist as an impractical dreamer.[56]

George Nelson, more of a purist, recalled disliking Dreyfuss for his commercial success reached without any "real far-out 'Bauhausy-type' designs." He seemed "immersed in the establishment . . . a terrible square . . . always taking the corporate side in any argument." But as he came to know Dreyfuss closely, he found in his work a solid "integrity." Dreyfuss, he realized, was "terribly interested in the relationship of his product to people."[57] He tried to make sure that products he designed reached an optimum level of fitness to the people who would use them. His office sought to "make machines fit people" instead of "squeezing people into machines."[58] His official credo recognized "that *what we are working on is going to be ridden in, sat upon, looked at, talked into, activated, operated, or in some way used by people individually or en masse*." Put more generally, "*If the point of contact between the product and the people becomes a point of friction, then the industrial designer has failed*."[59]

Dreyfuss settled on a more limited but more practical goal than either Teague or Geddes. Teague predicted creation of a golden age by application of design to society, and Geddes hoped to actualize his own imaginative projections, but Dreyfuss accepted society as it was. He considered each commission a discrete project and worked to ensure that consumers would find his product designs suited for their intended uses. Thus their lives would be improved, made more efficient and less irritating. His clients would benefit because he believed the public instinctively recognized good design. The aims of business and the public coincided through the industrial designer's mediation. Even Geddes, who criticized fellow designers Teague and Raymond Loewy, found no fault with Dreyfuss. He liked him: "I think he is honest, straight from the shoulder, there's nothing phoney about him, he's square—just a good human being."[60]

Unlike Teague and Geddes, Dreyfuss came from a background that directly influenced his later career.[61] He was born in 1904 into a New York family that had long operated a theater supply firm. After his father's death in 1915 destroyed the business, Dreyfuss helped support his mother and younger brother as a sign painter and delivery boy. In 1920 he entered Ethical Culture Arts High School on scholarship for the last two years of his education. The school was run by the Society for Ethical Culture, founded in 1876 by Felix Adler to separate ethics from theology. Since Adler believed society "an organism," he taught that "we live not merely or primarily to be happy, but to help on as far as we can the progress of things."[62] Ethical Culture schools therefore tried "to develop persons who will be competent to change their environment to greater conformity with moral ideals . . . to train reformers." Their graduates, Adler thought, should "believe that their salvation consists in *reacting beneficently upon their environment*"— with "an enthusiasm for progress" tempered by knowledge of the immense effort required for minor gains.[63] This conservative reformism reappeared in Dreyfuss's goal of improving the environment by fitting products to their users. Just as important for Dreyfuss was the school's art instruction, which emphasized "analogies between beauty and noble living." A knowledge of the arrangement of colors and shapes to produce harmony would stimulate a desire in students to form harmonious personal relationships.[64] Finally, the school provided Dreyfuss with practical experience in the profession he entered before becoming an industrial designer. To teach its students the art of cooperating in groups, the educational program included frequent pageants written, organized, and carried out by students. Dreyfuss designed sets and costumes for pageants with hundreds of participants.[65] After graduation in 1922, supported by an Ethical Culture scholarship, he entered a stage-design

class run by Geddes, then at the height of his theater career.

As a student and then for a year and a half as an assistant, Dreyfuss participated in some of Geddes's most ambitious theater projects while learning his craft. In 1923 producer Joseph Plunkett hired Dreyfuss to design sets for variety shows at the Strand Theatre, a job he held for four years, and in 1924 he joined the Roseland Amusement Company as a design consultant for its ballrooms in Manhattan and Brooklyn. A spell of unemployment in 1927, sparked by conversion of variety theaters into movie houses, sent the restless Dreyfuss to Europe. There he was tracked down by Oswald W. Knauth, a vice-president of Macy's department store, who had decided to offer him a five-figure salary as a stylist on the basis of his theatrical work. The job, as Knauth conceived it, would have Dreyfuss suggesting design changes for merchandise already sold in the store. His sketches would be submitted to manufacturers, who would have to change their products or find their business with Macy's terminated. Dreyfuss immediately packed his bags and took ship for a career in industrial design.

He spent two days in the store to determine changes that would modernize Macy's merchandise without costing much. His rejection of the post probably amazed Knauth. According to later accounts, Dreyfuss turned down the offer because he thought a stylist should work with a manufacturer at the inception of a product in order to gain complete understanding of limitations of materials and processes. Working through a department store, he could only provide cosmetic touches conceived in ignorance of factory conditions. He concluded that "an honest job of design should flow from the inside out, not from the outside in."[66] Equally important, Dreyfuss wanted to remain independent by acting as a design consultant to a variety of manufacturers instead of becoming an employee of one firm. In turning down Knauth's offer, he

39. Bell Telephone's first desk phone with speaker and microphone in a single hand-piece, 1927. From Art and the Machine *by Sheldon Cheney and Martha Candler Cheney. Copyright © 1936 by the McGraw-Hill Book Company, Inc. Used with permission of McGraw-Hill Book Company.*

40. Gustav Jensen. Model of desk phone proposed for Bell Telephone, c. 1929. Pencil Points.

had sought advice from Louis J. Brecker, a friend in the Roseland Amusement Company, who urged him to open his own office and introduced him to acquaintances whose products needed improvement.[67]

Although Dreyfuss continued to design for the stage until 1935, he was shifting gradually into industrial design. In 1928 he rented an office for twenty-five dollars a month and began executing commissions for such minor articles as snaps, buckles, and doorknobs. These jobs brought in little money and less publicity. When Dreyfuss moved to a new office on Fifth Avenue in 1929, he furnished it with a borrowed card table, folding chairs, and a twenty-five-cent philodendron. He could afford only two employees, secretary Rita Hart and business manager Doris Marks, a Vassar graduate whom he married the next year. Business was unpromising—primarily small things like hardware, plastic cigarette lighters, keys, watches, and canning jars. Most manufacturers who consulted him wanted a veneer of style applied to a product already designed by company engineers, but Dreyfuss's association with Bell Telephone Laboratories proved that he considered himself more than a decorator.

Bell's first desk phone with speaker and microphone in one hand-piece, introduced in 1927, looked awkward (see figure 39), even though engineers had based its design on the head measurements of four thousand people. Seeking a better design, Bell in 1929 offered a thousand dollars to each of ten designers recommended by the Art Center to sketch the ideal telephone. Despite the potential publicity of associating with a national corporation, Dreyfuss turned down the offer because he thought redesign of the phone should be conducted in collaboration with company engineers. A year later, admitting he was right, Bell hired Dreyfuss as a consultant on the desk phone project. Designs furnished by Bell's ten "commercial artists" had proven impractical because they failed to consider functional requirements.

Gustav Jensen's model of a set resembling the later Princess phone, for example, had a thin, straight handle, with six sharp angles running its length, which would have been uncomfortable to hold (see figure 40). Before the company retained Dreyfuss, its experience indicated that "the application of art to industrial objects" would fail until it overcame the "handicap" of artists with no knowledge of manufacturing. Dreyfuss's insistence on working with Bell's engineers made him seem just the man for the job. He began in 1930 to develop a new desk phone (figure 41). Introduced in 1937, it remained standard until 1950, when another Dreyfuss model replaced it.[68]

41. Dreyfuss. Desk phone for Bell Telephone, 1937. Courtesy of Bell Laboratories.

In his approach to style, Dreyfuss largely ignored European theorists like Le Corbusier and Mendelsohn. His version of the machine aesthetic seemed pragmatic and unprogrammatic. In 1932 he told a reporter that he aimed for simplicity, a quality that came to prominence with the skyscraper, "a straight, towering shaft with a simplicity that was unknown a few years ago in our buildings."[69] Many of his designs, such as the telephone, possessed simplicity of line without echoing the usual idioms. Others did reflect the modernistic or the streamlined style. But Dreyfuss neither contributed innovations to these modes nor accepted the conventions of other designers. With typically conservative restraint he used their motifs when appropriate and ignored them otherwise. His attitude derived not from ignorance but from a desire to fit a product's form to its function. A reporter summarized his attitude by concluding that he favored "simple *modern*" as opposed to "the garish and *modernistic*."[70]

Dreyfuss's central stylistic concern appeared in his designing of a theater in Davenport, Iowa, for the RKO chain in 1931. According to Dreyfuss, his design marked "an attempt to get away from all the over decorated auditoriums covering the country," a remark no doubt referring to lush Egyptian and Persian cinema palaces but per-

haps also to modernistic theaters.[71] Dreyfuss himself drew on the Paris style (see figure 42) but distinguished his work from its source. In a memo dictated for a publicity release, he described the Davenport theater as "in the modern style"—a phrase he qualified by asserting that "every effort has been made to infuse warmth into the entire scheme."[72] Apparently, he hoped to avoid both the cold lack of ornament of Europeans like Le Corbusier and Mendelsohn and the feverish exoticism of Paris. The quality of warmth, heightened by use of brown and gold colors, implied a middle ground in which the American movie-goer could feel comfortable.

Even in his work for RKO Dreyfuss demonstrated his concern for fitting designs, whether of products or buildings, to people who would have to use them. The company once sent him to Sioux City to see why a new theater was losing money to a dilapidated competitor. After lowering prices to no avail, Dreyfuss watched people walking past the theater for three days. This vigil convinced him that farmers feared tracking mud on a plush red carpet that ran from the lobby out to the sidewalk. When Dreyfuss replaced it with a plain rubber mat, the theater enjoyed full houses.[73] This anecdote suggests

42. Dreyfuss. Lighting fixture, RKO theater, Davenport, Iowa, 1931. Dreyfuss Archive.

Dreyfuss's attention to psychological nuances of the utmost importance to an industrial designer. Particularly in the twenties and thirties, when a machine age became conscious of itself through a machine style, carefully designed visual appearance became an important psychological factor in choices made by consumers. But Dreyfuss recognized that psychological considerations ran deeper than surface style. A layman's knowledge of human nature, reinforced by careful observation, could provide valuable assets to the design process. Dreyfuss's conception of industrial design did not include a scheme for restructuring the environment. Nor did he pursue publicity for himself and his profession. He dedicated himself instead to the practical task of simplifying the relationship between people and particular elements of their environment.

Making Life a Bit Easier

A native of Paris, Raymond Loewy (figure 43) was the only foreign-born among the major American industrial designers. During World War I, serving in the French army, he received several citations for bravery in combat. To relieve the tedium of trench warfare, he amused himself by creating a nonmilitary ambience. From ruined houses he liberated chairs, rugs, and a cracked mirror. On leave in Paris he cornered wallpaper, material for drapes, pillows, and copies of recent magazines, including *Vanity Fair*. Geraniums completed his temporary trench home, identified by a hand-painted sign as "Studio Rue De La Paix."[74]

Norman Bel Geddes once wrote that Loewy was "much more interested in living than in designing."[75] Another acquaintance of Loewy found his passion for life symbolized by his summer drives to Long Island, which included "the biggest white convertible and the most beautiful blondes." In essence, "he looked like a designer."[76] Eventually success brought him a villa on the

43. *Raymond Loewy posing with the 1939 Studebaker Champion and the Pennsylvania Railroad's S-1 locomotive, 1939. Raymond Loewy.*

Riviera, a seventeenth-century hunting lodge near Paris, a house in Mexico, another in Palm Springs, and an apartment in New York with a hinged Dufy painting concealing a recessed television. Pursuit of pleasure led him to deep-sea diving and, in his seventies, dune-buggy racing.[77] But Loewy never regarded design merely as a means of supporting a luxurious life style. His life and his work represented two sides of the same coin. Although he went into industrial design by chance, his motivation remained the creation of personally comfortable and tasteful surroundings. His career began with transformation of his wartime dugout.

Often Loewy has asserted that he personally "resents" ugliness and vulgarity. He expected to find America's environment composed of elements "simple, slender, silent, and fast." Instead he received an impression of "massiveness and coarseness." Gradually he realized that a subliminal feeling of "uneasiness" derived from "the actual form, sound, and color of things around me."[78] The longer he remained, the more "shocked" he was by the poor aesthetics of American machines, "by their clumsiness, how noisy they were, and vibrating, and unsophisticated." Life lost its sparkle in such surroundings. Even minor "parasitic factors" like leaky ball-point pens aroused Loewy's ire. As a designer his stated goal was "to improve things people live with from the moment they wake up till they go to bed" by eliminating "aggravations, irritations." Loewy's involvement in design rested on his personal harassment by ugly surroundings. By "making the average citizen's life a bit easier . . . , easier on his nerves," he fulfilled a desire to make his own life more pleasant.[79]

Loewy was born in 1893 into a typically bourgeois family.[80] As a child he exhibited interest in transportation machines and at seventeen entered a school devoted to preparing students for engineering school entrance exams. Although he never attended engineering school, three years of specialized instruction adequately prepared him for industrial design's technical aspects. Before the war Loewy worked a few months for an electrical engineering firm. Afterward, he could find no work in Paris and so left for New York in 1919, hoping to work for General Electric. On shipboard a chance event directed him to a far different career. To a charity auction held among passengers he donated a sketch of a fashionable young woman strolling on deck. Its purchaser was the British consul in New York, who admired his talent and provided letters of introduction to Rodman Wanamaker and Condé Nast, publisher of *Vogue*, to start the dapper Frenchman in a career as a fashion illustrator. As Loewy gained a reputation for having "a certain flair for sketches,"[81] he received assignments from Pierce-Arrow, the White Star Line, and Butterick. He designed costumes for Florenz Ziegfeld for a short time, but he concentrated on fashion illustration. In 1924 he met Henry Sell, editor of *Harper's Bazaar*, and soon was providing material for the magazine. Accompanied by fashion editor Lucille Buchanan, he went to dances, weddings, and other social functions to observe and sketch. When Saks Fifth Avenue opened in 1924, Loewy designed uniforms for elevator operators and ran an advertising campaign using "synthetization," defined as "isolating a single simple sales point in a sea of white space."[82] He continued illustrating advertisements for Saks and other Fifth Avenue shops until the end of 1928.

By that time Loewy had evolved a personal style distinct from typical modernistic illustration. In style and theme his advertisements exhibited a warm touch opposed to the cold precision of something like Geddes's Franklin Simon window displays. One of Loewy's Saks ads (figure 44) publicized a high-heeled shoe called the "Metropolis." A pile of abstract blocks rose to form an image of a skyscraper, superimposed on a dark sky punctured by bursts of starlight and cut by crisscrossing spotlight beams. At the pinnacle of this structure stood a leggy nude, supporting on outstretched hands the heel

44. Loewy. Advertisement for Saks Fifth Avenue
in Vogue, 1927. Saks Fifth Avenue.

and toe of a gigantic shoe, at which she gazed in rapture. This superstructure transformed the phallic skyscraper into an image of femininity. Rather than using sharp lines to delineate the form's abstract blocks, Loewy softened them by using changes in shading to indicate intersections of planes. Exaggerated perspective lent an air of unreality. Lettering in a rounded cursive script contributed a touch of femininity while appropriating the lazy-S of the jagged modernistic style.

In his fashion advertisements Loewy consciously reached toward sophisticated women who read *Vogue* and patronized Fifth Avenue merchants. But these illustrations also embodied elements of his own style—a warm, often flashy elegance, preference for soft tones, lighthearted wit, concern for detail, and preference for ornate typography. Faced with the streamlining and simplifying trends of the thirties, he submerged his preferences, but they did contribute to his industrial designs, becoming more prominent at the end of the decade in designs for a W. T. Grant store in Buffalo (figure 45)—a project that set the pace for his style throughout the forties and fifties. Uncomfortable with more sterile expressions of modernism, Loewy enlivened even restrained designs with a fillip of lighthearted elegance. Late in the forties he puzzled over the Coke bottle's success because it was neither streamlined nor stripped to essentials. But it was, as a reporter paraphrased his comment, "aggressively feminine—a quality that in merchandise, as in life, sometimes transcends functionalism." [83]

By Loewy's account, fashion illustration made him a wealthy man. Although he worked twelve-to-sixteen-hour days, he earned thirty thousand to forty thousand dollars a year—a comfortable sum considering that studio rent was his only overhead expense. As he recalled years later, however, he "got very restless" and wanted to do something "more meaningful." He "wished to return to engineering" but had no idea how to do it. Although he felt "aesthetically

shocked" by the vulgarity of American products and experienced "an irresistible urge to do something about it," he had no opportunity until "out of the clear blue autumn New York sky" in 1929 came Sigmund Gestetner, an Englishman with a duplicator (figure 46) to be modernized. [84]

Loewy had met Gestetner at a dinner in London and had impressed him with an analysis of the functional superiority of English taxis to those in New York. [85] Gestetner, now on a short visit to New York, wanted a new design to take home with him in five days. He had one of his mimeograph machines immediately sent to Loewy's studio. Using the machine itself as a form, the novice designer applied clay to show how he would "encase all the gadgety organs of the machine within a neat, well-shaped, and easily removable shell." [86] Previously, the duplicator's exposed gears and levers, saturated with ink, had collected dust and made operation messy. Loewy's shell (figure 47), eventually made of Bakelite, eliminated the machine's print-shop aura and lowered manufacturing costs by disposing of the nickel plating and hand-polishing of exposed parts. Loewy got rid of a dust-catching air space and replaced a boxy cabinet of lacquered sheet metal with a sleek wooden cabinet whose rounded horizontal edges were supposed to deflect dust. Short straight legs replaced long projecting legs that might have tripped the careless. His goal seemed to be simplification, both visual and functional (primarily ease of cleaning). [87]

The Gestetner job earned Loewy two thousand dollars for three days' work, gave him experience in designing products, and convinced him of industrial design's potential, but it brought him no publicity as a designer. Possibly owing to the Depression, Gestetner did not manufacture the design until 1933. But the experience did give Loewy confidence to seek other design commissions. As Loewy recalled, the first years were difficult, and he often felt a "dark, muddy despair." Handicapped by a prominent

45. Loewy. Interior of W. T. Grant store, Buffalo, 1939. Architectural Forum.

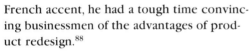

46. Gestetner duplicator, before redesign. Raymond Loewy.

47. Gestetner duplicator, after redesign by Loewy, 1929; introduced 1933. Raymond Loewy.

French accent, he had a tough time convincing businessmen of the advantages of product redesign.[88]

About 1929 Loewy became an art director for Westinghouse in charge of radio cabinet design, a position he held until at least 1931. Despite his condemnation of skyscraper furniture as "monstrous," he contributed at least one design to the genre, a slim wooden radio cabinet.[89] More significant was his association with Shelton Looms, a textile company. In the late twenties he created some fabric designs for the company and edited its house organ *The Weaver*, which took first place in a national competition. With architect Ely Jacques Kahn he collaborated on new interiors for the firm's office, opened in June 1929. One contemporary account described their work as "in the style of a modernistic living room."[90] This association with Shelton led to employment with the Hupp Motor Company as a consultant on textiles for upholstery, a position he par-

layed into being assigned the design of an automobile.[91]

Loewy had already thought about automobile design. In 1928 he filed for design patents covering two bodies, a headlight, and a radiator. Sleek and low, with abbreviated running boards and slanted windshields and window posts, both bodies exhibited continuity of line between hood and body proper.[92] Hupp hired Loewy as a body designer after seeing these designs, but nothing significant resulted for several years. Although Jack Mitchell of Hupp's advertising agency supported Loewy, he experienced difficulty convincing the firm's engineers that his ideas were practical. His contributions suffered radical transformations in final production plans. Eventually he won over engineers with a privately built body on a Hupmobile chassis completed in 1931 at a personal cost of eighteen thousand dollars. After this prototype won styling awards, the company gave him a greater part in the design of the

48. Loewy Model of 1934 Hupmobile. Raymond Loewy.

1934 Hupmobile (figure 48), begun in 1932.[93] This association gave Loewy a glimpse of the frustrations industrial designers were to experience in dealing with engineers protective of their own traditional prerogatives. More important, it opened to him the field of transportation design, for which his office would become best known in the thirties.

In 1929, however, prospects looked grim to Loewy. He had completed one major design, the Gestetner mimeograph machine, which had not gone into production. His relationship with Hupp had possibilities but had so far proved unsatisfying. To top everything off, the stock market crash swept away his savings from fashion illustrating—a career he had given up for the uncertainties of industrial design. In the following general depression in confidence, he lost most of the few design accounts he had managed to attract. Loewy and the other fledgling designers who had ridden the final nervous

crest of the prosperous twenties could not foresee, of course, that in a few years manufacturers would be beating down their doors for help in reversing plummeting sales curves. Or that they would respond with a gleaming streamlined style, purged of modernistic excesses and calculated to restore an optimistic faith in big business and its technological innovations.

Selling Industrial Design

... in the future we are going to think less about the producer and more about the consumer.
— *Franklin D. Roosevelt*

Industrial designers have always considered their profession a "depression baby."[1] Even though advertising agents and department store executives took product styling seriously two years before the crash, it did not concern most manufacturers until after full economic collapse. They turned to radical solutions because they had nothing to lose; to the optimistic or the desperate it seemed a panacea. Within a few years industrial design became a fad among manufacturers. In 1933 an executive whose profits doubled after Joseph Sinel restyled a hearing aid declared that a "redesign movement" would solve the nation's economic problems. Thousands of toolmakers and die-makers would find work, steel mills would increase output to supply new machine tools, pulp mills would answer demands for new packaging, and the need for raw materials would radiate "in a thousand directions." As a result the economy would generate "wages and more wages—to be spent for the new, the desirable, the irresistible product." This scenario seemed a "logical prediction."[2] Even more optimistic, designer Geddes declared in 1932 that historians fifty years hence would have to explain why, "in the midst of a world-wide melancholy owing to an economic depression, a new age dawned with invigorating conceptions and the horizon lifted."[3]

Early in the Depression, however, industrial designers faced horizons as limited as anyone else's. From 1929 to 1932 net income from manufacturing in the United States fell by more than two-thirds. Recovery began in 1933, but income did not approach the pre-Depression level until 1937.[4] Even manufacturers who stayed in the black could not at first afford to plow limited profits into retooling. In addition, outside consultants were among the first luxuries eliminated when businessmen retrenched. Although industrial designers launched their careers to answer a demand for distinctive products in a buyer's market—a need heightened after the collapse—they found little or no work.

Loewy retained the Hupp Motor Company as a client but received few other commissions until the mid-thirties. Teague struggled along with three or four accounts until 1933, when his firm served eleven clients. The following year he had seventeen. Dreyfuss felt the crunch most. His list of thirty-eight clients in 1930 fell to sixteen the following year, a figure that remained stable through the decade. And Geddes, who had never had many clients, limped along with one or two jobs a year until 1934, when his office boasted eight clients.[5] To succeed, industrial designers had to convince manufacturers that they could contribute to economic recovery.

Fortunately, the advertising men who had promoted them in better times furthered the selling of industrial design with a general rationale calculated to evoke gut response from businessmen. Largely a product of the Calkins and Holden Agency, the scheme, like most others for national recovery, attempted to deal with the overproduction-under-consumption dilemma through rational planning and control of the economy.

The Consumer Today Is King

As Joseph Dorfman noted in his history of American economic thought, proposals for coordinating production and consumption during the Depression differed with "the divergent backgrounds of their sponsors."[6] Observers did not even agree on reasons for the imbalance. In an ominous article, "Obsolete Man," *Fortune* argued that, despite abundance, thousands of workers had lost their jobs to machines as manufacturers took advantage of technological innovation. No longer needed in production, these workers were unfortunately not obsolete "from the consuming point of view," but they lacked purchasing power.[7] Studies by the Brookings Institution, on the other hand, questioned the premise of abundance itself. Even maximum use of the nation's factories would raise everyone's standard of living only

slightly and could not satisfy "vast potential demands" for "basic commodities" and "conventional necessities."[8] This Brookings analysis demonstrated an unusual social concern. More often those who proposed solutions to the economic crisis tried to tinker with the production-consumption machine, to remove the bugs preventing efficient operation without considering the fate of people who had never benefited from it in the first place. The image of an engineer achieving economic control by applying scientific techniques appealed to individuals across the political spectrum. Radical Howard Scott's briefly popular Technocracy movement, for example, advocated replacing the capitalist pricing system with a distribution system in which each citizen would receive an equal number of "energy certificates" redeemable for goods priced according to the amount of energy used in making them.[9] But the engineering mystique also attracted conservative individuals like adman Earnest Elmo Calkins.

In 1930 Calkins contributed with other businessmen to a symposium on the "philosophy of production." The editor, J. George Frederick, was disappointed that America's manufacturing system could "click off" products as fast as a printing press turned out newspapers but was "compelled to stop its fascinating wheels and shelve new scientific discoveries because nobody knows what to do with the heaped-up product."[10] Responses to the problem varied. Some contributors addressed the overproduction horn of the dilemma. The chairman of the board of General Electric, Owen D. Young, suggested dumping surpluses abroad, but he also thought manufacturers should act together to limit production in order to avoid government regulation.[11] Bernard Baruch, who had coordinated industrial output as head of the War Industries Board, echoed this solution. Businessmen had no alternative to providing the "national producing engine" with a "governor" by acting together to limit production.[12]

Other symposium contributors considered the economic crisis a matter of underconsumption rather than overproduction. Businessmen who hoped to win out as individuals in a competitive market rejected the idea of regulation, whether by industry or government. In general, business viewed the crisis as a matter of underconsumption brought on by the refusal of individuals to spend money that instead went into savings. This belief relieved businessmen of responsibility for the Depression even if it did not relieve the pressure. Henry Ford attacked consumers for thriftily maintaining old goods instead of replacing them with more advanced products. Businessmen, he thought, must create demand for an ever-increasing flow of goods that would boost purchasing power through increased wages.[13] Charles F. Abbott of the American Institute of Steel exhorted readers "to do more *creative wasting*, in order to gain more progress." Abbott advocated *"remaking our whole world"* through a program of "progressive obsolescence," which would require "a more statesmanlike view of selling and advertising" to keep factories running.[14] Any attempt to brake America's production machine would, in the words of editor Frederick, have the impact of "a child playing on the track of the Twentieth Century Limited."[15]

Earnest Elmo Calkins adopted the underconsumption position in his contribution to the symposium, "The New Consumption Engineer, and the Artist," an opening salvo in his agency's promotion of a coherent business rationale for industrial design. Calkins argued that the American public's capacity for consumption was limitless. But factories lay idle because businessmen continued to supply products that did not fulfill true needs of the public. In the twenties they had relied on high-pressure sales tactics to push unwanted goods, but the fever pitch of this artificial stimulation could not be maintained. In order to balance production and consumption, business had to turn to "consumption engineers," who would "find out

what is clogging the flow of goods and remove the obstacle." Calkins considered consumption engineering a "new business science," which would enable industry "to dig in deeper and anticipate wants and desires not yet realized, but foreshadowed by trends and implicit in the habits and folkways." Using market surveys, consumer questionnaires, and behavioral psychology, consumption engineers would predict changes in buying habits and end disastrous trial-and-error marketing. Beyond this, they would truly engineer consumption by manufacturing needs that had not before existed. As examples of this function, Calkins cited two contradictory projects—his own support of a national highway system to promote auto sales and Edward A. Filene's naive attempt to stimulate shoe sales by promoting the joy of walking. A relic of a genteel age, Calkins did not imagine the future subtlety of consumption engineering.[16]

Behind his proposal lay a reversal of "the garnered economic wisdom of the centuries." He argued that "prosperity lies in spending, not in saving." Calkins divided goods into two groups—"those we *use*" and "those we *use up*." Consumption engineering would "help us use up the kind of goods we now merely use." Although he advocated easy credit, high wages, and employee profit-sharing plans to prime the pump, these were afterthoughts. His program's central tenet was the creation of "artificial obsolescence" through the redesign of products, "something entirely apart from any mechanical improvement, to make them markedly new, and encourage new buying, exactly as the fashion designers make skirts longer so you can no longer be happy with your short ones." Such examples as Teague's cameras and Geddes's metal beds—demonstrating "a pleasing working out of the envelopes of the product"—convinced potential consumers that they could not be stylish without possessing them. In conclusion, as if consumption engineering seemed a bit crass, Calkins asserted that mass-produced objects of taste and beauty would improve public taste and relieve "the ugliness and spiritual poverty of much of this modern machine environment."[17]

Second thoughts did not occur to Egmont Arens, director of the Industrial Styling Division of Calkins and Holden, who spent several years developing his boss's remarks, proselytizing among manufacturers for a program he referred to more smoothly as "consumer engineering." He hoped not only to stimulate a demand for designers' services but also to gain packaging and design clients for his own agency. After writing a number of articles on the subject for trade journals, Arens collaborated with Roy Sheldon, another Calkins and Holden packager, on *Consumer Engineering: A New Technique for Prosperity*. They developed the basic concept beyond the level of sophistication of Calkins's original proposal. Their mentor, however, set the tone of their discussion by introducing consumer engineering as a technique for eliminating "the obstacles in the way of the free flow of goods from factories to consumers."[18] The consumer engineer's job, in other words, was to eliminate friction in the distribution machine—not by redesigning the machine as the engineers of Technocracy suggested but by applying lubrication.

American business had benefited earlier from one efficiency movement. As Sheldon and Arens noted, "production engineering" had "greatly increased the effectiveness of the individual in production."[19] They were referring to a factory management fad that swept industry after the publication in 1911 of Frederick Winslow Taylor's *Principles of Scientific Management*. An engineer, Taylor had tried to organize production scientifically by analyzing each worker's task and rearranging its components into a sequence that would be less wasteful, though often more boring. His approach made machines of workers.[20] In the twenties, however, by exploring subtle factors of employee motivation and morale, industrial psychologists

extended management techniques from abstract jobs to actual workers.[21] To Sheldon and Arens it appeared that factory management had rendered the American production machine so efficient that its output was going to waste. The answer lay not in limiting production but in rendering distribution equally efficient.

The twenties had witnessed the "Taylorist" phase of consumer engineering as merchandisers focused on the mechanical job of distribution. Mass national advertising, installment buying, chain stores, and style rotation plans had rendered distribution efficient from the manufacturer's point of view. But these developments were insufficient. Just as factory management had eventually turned to psychology to illuminate and control the behavior of workers, consumer engineers would have to discover how "to make each of us a more active and efficient consumer." It would have to concern itself with "states of mind" to solve the problem of "how to increase our receptivity, how to stimulate and create desires, how to overcome antagonisms and inertia."[22]

For Sheldon and Arens consumer engineering involved two steps. First, questionnaires aimed at both dealers and consumers would reveal product qualities that appealed most to potential buyers. Second, designers would embody those qualities in restyling. Both steps involved "humaneering," a euphemism for psychological manipulation. Inspired by John Dewey, who had, they claimed, supported expansion of the scientific method to social affairs, Sheldon and Arens looked to psychologists to provide "a human technology to solve the human problem."[23] They knew that behaviorist John B. Watson had been working at the J. Walter Thompson Agency for more than ten years as a result of Stanley Resor's desire for an associate who could "codify the laws of human reactions."[24] According to Watson, the psychologist in merchandising studied the buying habits of "Mr. and Mrs. Consumer" as closely as "habits of mice and monkeys in the well known experiments of academic psychologists."[25] By advocating his reductionist approach, Sheldon and Arens subordinated expressed needs of people to a desire to get the distribution machine running smoothly at any cost. The consumer engineer applied the concepts of psychology "not only to fit the product and the promotion to the existing market, . . . but to create new needs and stimulate consumption by every possible means." They hoped for total control of the consumer, who would become just as much a cog in the distribution machine as the worker in the production machine.[26]

Their goal was artificial obsolescence, expressed picturesquely as "progressive waste, creative waste."[27] Recently historian Donald J. Bush has suggested that Sheldon and Arens relied on technological innovation to fuel obsolescence.[28] The two admen did in fact note the importance of technological obsolescence, but they considered consumer engineering primarily a technique for differentiating a product from its competitors after all brands had reach a plateau of mechanical efficiency.[29] Although the public reacted with disgust to negative connotations of "obsolescence," they argued that people actually hungered for goods expressing a modern spirit in the latest styles. It is ironic to find an efficiency movement engineering waste, but Sheldon and Arens were more concerned with distribution as a process than with its presumed goal of satisfying society's material needs. Their program would, they argued, "reduce waste to a minimum" by eliminating "blind guess and hunch" from merchandising. True "waste" in the economic system derived not from the engineered replacement of serviceable goods but from manufacturers who stubbornly refused to embrace redesign. If such laggards hoped to reverse plunging sales curves, they would have to consult the consumer engineer, "a Luther Burbank of the sales-counter, who can make two sales grow where only one grew before; a Mussolini, or,

if you prefer, a Lenin, . . . forever conspiring and effecting silent and bloodless revolutions in the habits of nations."[30]

Geddes once observed that vehicles were "streamlined in order to *eliminate* disturbances in the media through which they pass."[31] Discussions of the underconsumption problem relied on similar terminology. The popular "flow of goods" metaphor paralleled the flow of a vehicle through air. The authors of *High-Level Consumption* compared this flow of goods to "the resistless current of a mighty river" but noted that its passage could be interrupted by "whirlpools, eddies and waves"[32]—just as projections on the surface of an auto created slowing eddies of air. This aerodynamic "parasite drag" had its counterpart in the "psychological drag" of consumers who refused to give up the old for the new.[33] Like air resistance, sales resistance had to be eliminated. Marketing's goal, according to a psychologist, was not to break down consumer resistance but "the discovery of how to avoid sales resistance, how to discover and sell articles to which there will be the least resistance."[34] The ultimate use of this metaphor occurred in an article written to boost industrial design after the streamlined style had become widespread: "Streamlining a thing strips it for action, throws off impediments to progress," but more important, "streamlining a product and its methods of merchandising is bound to propel it quicker and more profitably through the channels of sales resistance."[35]

Such considerations evolved from Sheldon and Arens's vision of an ideal society as a frictionless production-consumption machine engaged in continuous elimination of free-flowing waste. This vision degraded people by regarding them as binding points of friction to be eased with a dose of the oil of psychological manipulation, yet it appealed to businessmen in need of economic relief. It reinforced the morale of designers, who experienced in the early thirties, as a Geddes associate recalled, "a jubilant feeling that industrial design was going to revolutionize industry, that it would become as big and as independent a force in the economy as advertising."[36] And those designers inclined to consider crass the purposes of consumer engineering could take to heart Calkins's optimistic assertion that it would "satisfy the soul" by shaping a harmonious material environment and simultaneously relieve the Depression.[37] Calkins's positive rationale appealed to designers criticized for the obvious commercial intent of consumer engineering. To carpers who claimed that industrial design ran counter to the "rugged and serious thinking of the Depression" by stimulating a "yen in us for change, for going places, for getting there," with no goal orientation,[38] they answered that they were creating a machine civilization free of functional inefficiency and visual incongruity. But first came the task of selling their services to manufacturers.

Profits from Depression

Earlier than most designers Geddes recognized the need for publicizing his practice among manufacturers. By 1944, when he issued his standard practice manual for employees, he had developed a systematic method for publicity, in theory at least. Before he solicited a company's business, his public relations department placed an article in a relevant trade journal. Ostensibly about design in general, it would establish Geddes as an authority in the particular field. If a potential client responded with interest, the office prepared a description of services it could provide, accompanied by articles in general circulation magazines and newspapers with favorable mention of Geddes's work. Finally, if negotiations reached the contract stage, the public relations department arranged publication of a small news item concerning Geddes, which could be a social or theatrical notice. In addition to attracting clients, articles ghostwritten for

Geddes contributed "a large income" to the office.[39]

Efforts to publicize industrial design early in the Depression hardly reached this complex level. But designers and their advocates realized that to many manufacturers, paralyzed by economic collapse, tinkering with a product's appearance seemed like "fiddling while Rome burns." They had to be shown how redesign would work for them. They had to be convinced, as one publicist claimed, that "to put up a well designed front is to put up a brave front and is . . . one of the ways to beat Old Man Depression."[40] The resulting spate of articles on design revealed not only the competitive advantages that designers hoped to provide clients but also the prejudices and hostilities they had to overcome.

Much early publicity for industrial design emanated from advertising agencies, appearing in such journals as *Printers' Ink*, *Advertising and Selling*, and the latter's graphics supplement, *Advertising Arts*—which in the absence of a specific design journal for a few years provided a forum for designers. Calkins and Holden led the way with its Industrial Styling Division and consumer engineering, but it was not alone. In an address to the Art Directors Club delivered in April 1930, Vaughn Flannery, art director at Young and Rubicam, exhorted colleagues "to assist in the promotion of industrial design" by acting as "liaisons" between designers and manufacturers.[41] Geddes, Dreyfuss, and Loewy had all previously benefited from such matchmaking as admen sought to convince businessmen to modernize their products. Admen considered industrial design a subsidiary of advertising. Convinced of the folly of using high-pressure techniques to push undistinguished goods in a saturated market, advertising agencies hoped to relieve pressure on them to produce sales by passing the buck back to manufacturers. If they hired stylists to make products more attractive, then agencies might have some chance of mounting effective campaigns. As an

49. Dreyfuss. Toaster for Birtman Electric Co.,
1933. Reprinted by permission of Modern
Plastics Magazine, McGraw-Hill, Inc.

executive of Lord and Thomas phrased it, "what the advertising business has to have from the designer is principally a means of dramatising and presenting a selling idea."[42]

The design propaganda of advertising agencies during the early Depression continued to maintain the upbeat rhythms of the new American tempo. The Lyddon, Hanford and Kimball Agency offered manufacturers a free pamphlet by partner Abbott Kimball, "Going Modern." According to an announcement in *Advertising Arts*, the brochure addressed the question "What shall we do about this modern style?" Kimball revealed how to "change the appearance of merchandise; the character of products; the form of containers and cartons" to bring them in step with the "new tempo" and its "tall angular women . . . metallic planes . . .

new gamut of colors . . . strange forms and abstract figures."[43] But as Kimball himself noted a year later, in 1931, feverish modernistic design seemed inappropriate to a Depression state of mind. In a time of "kaleidoscopic impressions"—a euphemism for social and economic chaos—simplification was the only way to "make any impression upon the human mind."[44] Neither admen nor designers heeded his advice for the moment. Advertisements continued to reflect the predominant style of the twenties, while designers began to create mass-market goods in an idiom once limited to luxury objects—producing gaudy designs like a stepped clock with projecting sun rays, a multifaceted coffee pot, and a "cubist" toaster, the latter by Dreyfuss (figure 49).[45]

Such extravagant designs did not fare well

at the hands of consumers. By 1934 recognized designers like Joseph Sinel were calling their own profession a racket run by unqualified stylists ignorant of manufacturing processes and supported by admen who advanced industrial design as a panacea because they were unable to do their own jobs.[46] Conditioned by the prosperity of the twenties to think they could work miracles, admen assumed they could force acceptance of anything as long as it was radically new and bizarre. But manufacturers tended to distrust glib salesmen proven inadequate by the Depression. As a result, industrial design seemed suspect to the extent that it was identified with advertising. Design publicity in advertising journals thus had more influence with admen and designers than with manufacturers. By 1934 even *Advertising Arts* could argue that industrial design would become "a concomitant of the New Deal—a new and vital force of great creative potentiality" only by renouncing "that type of redesign someone has called face-lifting." There was "no future for it as a by-product of selling technique."[47]

More significant in winning manufacturers than publicity in advertising journals were articles on industrial design in general business magazines. Early in 1930 *The Business Week* proclaimed "The Eyes Have It." According to this article, "stylizing" provided "an effective weapon in meeting new competition." Aware of business distrust of design, the article claimed that the contemporary Style Advisor or Art Counsellor provided style instead of the effeminate "fashion" of a few years earlier. "Modernistic angles and jumbled lines" had yielded to "something more subdued, simple, practical." Although not accurate at the time, this assertion inspired confidence. A *Forbes* writer announced that "modern design is emerging from the fad stage, and the demand for it is developing into a well-established trend," ignored by manufacturers at their peril. Unlike *The Business Week*, which had presented brief accounts of specific de-

signers, *Forbes* merely described several successful designs to emphasize consumers' acceptance of the modern style.[48] Owing to mundane graphics, such magazines hardly rendered industrial design in its best light. More successful was *Fortune*, the most energetic champion of redesign among general business magazines.

An idealistic article about Geddes, published in July 1930, downplayed the importance of profits in design. Instead, *Fortune* emphasized his contribution to a "modern estheticism" that would "embrace the machine with all its innuendos." While criticizing the outmoded design sense of most businessmen, Geddes—according to *Fortune*—took pains to distinguish himself from the long-haired artist—the bête noire of businessmen contemplating design. An executive might concern himself solely with profits, but a designer like Geddes "will examine his product with detached, speculative gaze, will judge of its requisites and possibilities abstractedly, and very probably re-create it in a new, more utile, more viable form." In his designs Geddes maximized beauty and utility, leaving profits to follow according to "the proposition that the greatest profit may be attendant upon beauty, that whatever is stupid or ugly cannot possibly be functionally superior."[49] To accompany the text *Fortune* reproduced Al Leidenfrost's renderings of Toledo Scale's factory and scales, a rendering of a Simmons bed, and two photographs of the J. Walter Thompson room touched up with textured charcoal to simulate Leidenfrost's mystical style. Printed in dark burnt orange, these illustrations contained white areas so bright that contours seemed burned away by an intense illumination radiating from within the objects.

As the decade progressed, *Fortune* continued to publicize design but addressed more pointedly the professed needs of businessmen. This direct approach characterized a full-scale review of industrial design that tried in 1934 to dispel misconceptions that had hampered executives' acceptance of design. The author concurred with most observers in describing industrial design as a "depression-weaned vocation" promoted by advertising agencies after it became obvious that "the product had to be made to sell itself." He (or she) warned potential clients that the profession was "filled with pretenders and visionaries" who ignored cost and utility in creating extravagant designs. Reputable design, on the other hand, possessed a "three dimensional quality" derived from consideration of "the appearance of the article, the cost of the article, and its usefulness." The author asserted that "whenever two products are equal in point of utility and price, the one that looks most attractive to the purchaser will be bought first." But while improving a product's appearance, a reputable designer inevitably considered function and manufacturing cost. Thus, *Fortune*'s concept of industrial design went beyond mere styling and appealed to manufacturers worried about placing their success in the hands of superficial admen.[50]

For manufacturers in search of hard facts *Fortune* provided biographical sketches of ten designers and eight case studies demonstrating that costs of retooling were often offset by lowered production costs, and that sales might rise anywhere from 25 to 900 percent after redesign. Apparently, even in the midst of the Depression beauty did, indeed, pay. To drive home this point, *Fortune* accompanied the article with before-and-after-redesign photographs of a toaster, a check register, and a stove, all of which had benefited considerably. Most revealing was a shot of Lurelle Guild's "waiting room," a junk-store clutter of lamps, appliances, and kitchen utensils ready for face-lifting.

There is no way to measure the impact on manufacturers of articles like this one. But it did give industrial design the imprimatur of the nation's leading business magazine. At least one executive sought Dreyfuss's services after reading about him in *Fortune*.[51]

There must have been similar cases. But general articles did not attract as much notice as those in trade journals published for specific industries or particular occupations—such as sales managers or product engineers—in which designers and their fellow travelers could focus attention on concrete concerns.

Often designers themselves lectured executives on sales advantages of industrial design. Their articles yielded free publicity as well as an author's fee. Teague, for example, told readers of *Glass Packer* that designing glass bottles solely for ease of manufacture and filling ignored the important fact that bottles were not "bought by machines" but "by women."[52] And according to Dreyfuss, writing in *The Coin Machine Journal*, a vending machine suffered "the handicap of silence" because it had to "attract the customer and then sell its wares." Since battered, filthy machines repelled customers, he advocated upgrading the vending machine's image by replacing old ones with more attractive redesigned models, by installing machines in apartment house lobbies, and by instituting "a constructive program telling mothers that their children are safe in buying in this way."[53] The last two suggestions contributed a disinterested tone to his plea while highlighting his freewheeling inventiveness.

Testimonials by manufacturers also belonged in the public relations repertoire of industrial design. Articles by them often revealed steps taken in the design process to impress executives with its rationality. Howard E. Blood, president of the Norge Corporation, contributed articles to various journals praising design in general and Lurelle Guild in particular for the success of a new refrigerator. After analyzing a hundred thousand entries submitted in a contest requiring suggestions on improving refrigerators, the firm had told Guild to design a cabinet that would not look like a bulky icebox. The designer recessed the door to bring it flush with its frame, got rid of heavy hardware, rounded off corners and edges, and extended the base to the floor. As Blood phrased it, the cabinet exhibited "the sheer verticals and rounded horizontals of modern simplicity." The redesign job had quickly "jumped volume beyond expectation." Industrial design, he concluded, would contribute to "the comeback process" when designers worked from the solid findings of market research.[54]

More frequent than testimonials were "objective" case studies. Although designers often worked hard to place these notices in trade journals, they exuded an air of independent authority. One such article printed by *Sales Management* in 1936 provided a wry twist to the Norge story. Confronted by growing popularity of electric refrigerators as prices fell, the nation's icebox industry had watched its sales decline from 1,500,000 in 1924 to 225,000 in 1932. But New Hampshire's Maine Manufacturing Company, in business since 1874, refused to give up. Instead, the firm hired designer Robert Heller, who incorporated innovative lightweight insulation in a rounded sheet-metal cabinet with a thin, unobtrusive door. Thus, the icebox became the "modern ice refrigerator" and enjoyed a rise in sales of 100 percent. If redesign could salvage even obsolete products, the article implied, think what it could do for modern ones.[55]

Often the case-study approach led to regular columns describing four or five successful designs and providing brief accounts of design processes, design and manufacturing innovations, sales figures, and photographs. These columns appeared most frequently during the transitional years of 1933 and 1934, when industrial design was changing from a suspect fad to common practice. Thomas J. Maloney's "Case Histories" appeared in *Advertising Arts* every other month in 1934, and his "Case Histories in Product Design," sometimes a collaboration with designer George Switzer, appeared monthly in *Product Engineering* through much of 1933 and 1934. Although

these columns disappeared as the need for constant publicity faded, *Sales Management* maintained its monthly "Designing to Sell" column from 1932 through the decade's end. The brief case studies in the columns often pointed out cooperation between designers and product engineers. An ironing machine designed by Harold Van Doren and John Gordon Rideout, for example, boasted engineering by P. Euard Geldhof of the Easy Washing Machine Corporation. And the engineers of the Quiet May oil burner (figure 50) had compressed formerly clumsy controls and pump in a compact unit easily enclosed by Lurelle Guild in a streamlined housing appropriate for a basement "renovated to serve as a play room or bar room." Such overtures to product engineers hinted at the hardest battle in selling industrial design—convincing veterans of manufacturers' product design departments that industrial designers were not, as Dreyfuss recalled common opinion of the early years, "interlopers invading an area sacrosanct to the engineering mind."[56]

Traditionally, the same engineers had designed products and coordinated production. Recently, however, product engineering, which focused on functional arrangement of a product's mechanical parts, had evolved from production engineering, which directed the manufacturing process. With jobs as uncertain for engineers as for anyone else, members of this new occupation felt threatened by industrial designers and attacked them as impractical artists. Product engineers could handle redesign without relying on outside consultants of dubious competence. Early in the thirties *Product Engineering* therefore undertook to educate its readers in design for appearance. "The Product Needs Color," they learned, because it stimulates consumers' emotional responses.[57] New products like radios should rely on natural expressiveness of materials to avoid the extra expense of making products "look like what they are not."[58] Product engineers should emulate colleagues in the

50. May Oil Burner Corporation pump and controls, before (below) and after (above) redesign by Lurelle Guild, 1934. From Art and the Machine *by Sheldon Cheney and Martha Candler Cheney. Copyright © 1936 by the McGraw-Hill Book Company, Inc. Used with permission of McGraw-Hill Book Company.*

appliance division of Thomas A. Edison, Inc., who had recognized the significance of "questions of external appearance" by providing a sandwich grill with a "modernistic" Bakelite handle and designing a coffee pot according to the formulae of dynamic symmetry.[59] Such articles received support from editor Kenneth H. Condit, who frequently noted engineers' responsibility to stimulate sales during the Depression by rushing improved designs into production. The product engineer had to "drive himself to the limit" to keep factory workers employed.[60]

Not many product engineers listened. Condit found in September 1931 that only 7 percent of his readers considered appearance a major factor in product design. He exhorted them to adopt redesign if they wanted to avoid replacement by some "artist."[61] In the same month his managing editor, George S. Brady, in an article in *Mechanical Engineering* warned engineers to consider public desires as well as technical details when designing products if they wanted to satisfy executives apt to hold them responsible for sales. The product engineer who refused to yield risked being replaced by "an outside artist, or even a sculptor."[62] To emphasize contempt for such creatures, *Product Engineering*'s editorial team refused even to use the term "industrial designer."

This common hostility prevented hesitant manufacturers from hiring consultant designers. Even when hired, an industrial designer had "to gain the sympathetic cooperation of the engineers" before he could accomplish anything.[63] By June 1932 Condit had realized that artists were not going to replace engineers and that, in fact, they could provide aesthetic qualities that most engineers were not trained to provide. He ran an article by "design consultant" Teague concerning the need for "team work" between artist and engineer. According to Teague, an industrial artist shunned "the merely visionary" and had to "understand machines and techniques, and have a wide

knowledge of production methods." To create a successful design he had to work "in the factory where it is to be produced" and engage in a give-and-take process with product engineers. Rather than replacing engineers, he supplemented their expertise, bringing from his "broad and diversified experience" a valuable sense of style—"the way people in general at a given time like things to look."[64] Capitulating completely, managing editor Brady declared a few months later that the "use of outside consultants for criticism and suggestion has become general," but cooperation with engineers remained essential.[65]

By the end of 1932 Condit felt secure running a conciliatory editorial, "Engineer and Artist," in which he reported conversations with two "industrial designers"—a term finally admitted to his pages in quotation marks. One lamented the obtuseness of manufacturers who forced him to work with sales departments rather than with engineers, who thus remained suspicious. He insisted that he was not competing with engineers because "it takes one type of mind and training to design a smoothly operating machine and another to give it a presentable, salable exterior."[66] Other designers echoed this rationale. Addressing the American Society of Mechanical Engineers, Joseph Sinel stated that the designer's goal was "to improve the optical aspects of purely functional parts" devised by product engineers.[67] Designer Otto Kuhler, who advocated streamlining trains to attract passengers despite a belief that streamlining served no aerodynamic function, asserted that a "product is more salable if technical perfection is expressed by a well-designed outside appearance." A mathematically determined form—a ball bearing or a suspension bridge—possessed intrinsic beauty, but most products needed a "trained artist . . . to smooth the lines which technical convenience has given a product and emphasize by contrasts in color or material the inherent beauty of any utilitarian product, so as to make it *look* compact, durable, efficient, or whatever attribute might be necessary to improve its marketability."[68]

In selling their profession to product engineers, industrial designers renounced the "fog and glamor" of their advertising connection. They admitted a need for familiarity with manufacturing processes and the possibilities of various materials. And they accepted collaboration with permanent engineering staffs. They had to talk shop with engineers, but in reality they focused on surface qualities of products whose functional designs were completed. They worked with engineers, but their primary responsibility was to a firm's sales department. When under attack for superficiality, however, they played the engineering connection for all it was worth. Publicity for the general public always emphasized improved appearance but usually included mention of functional improvement of products. Such juggling led to contradictions. Teague's discussion in *Product Engineering* of style as transient, for example, contradicted a frequent assertion that he was engaged in discovering an ideal form for each type of product. Such incongruities developed as designers explained themselves piecemeal to groups with diverse interests: sales executives, product engineers, and consumers.

Problems of definition did not confuse relations between designers and the plastics industry. A corporate designer claimed in 1935 that industrial design owed its success to the rise of plastic as an industrial material.[69] This statement contained a grain of truth. Plastics constituted a growth industry in the Depression's economic wasteland. In "Plastic Southern California," a Los Angeles custom-molder boasted that use of plastics was growing as fast as his city. He found himself besieged by requests for, among other things, "the molded devices the doctors want in their efforts to keep Hollywood pure."[70] This dubious joke revealed the industry's only problem. Manufacturers and public alike considered plastic sleazy. The

twenties had witnessed cheap buttons and jewelry, marbleized combs, knickknacks with grotesque molded "carving," and items too obviously disguised as wood. Plastic had a reputation "as a cheaper substitute."[71] To move plastic beyond a limited novelty market, the industry publicized the material's inherent beauty. In other words, it embraced industrial design. Plastics and industrial design enjoyed a symbiotic relationship.

In 1930 *Plastics and Molded Products*, the industry's chief journal, concentrated on technical articles describing new types of plastic and new molding processes. Occasionally it ran articles boosting plastic's design value, illustrated with small, poorly reproduced product photographs. In August the journal highlighted three clock cases rendered "appealing to the eye" with Bakelite.[72] And in December it announced: "Christmas Gifts for All, of Synthetic Plastic Materials Are Attractive, Appealing and Acceptable." Most illustrated products betrayed no hint of modernistic design, but they demonstrated the variety of colors and forms possible with plastic. A marbleized brush set, according to a blurb, "almost makes one lose faith in the natural, so handsome is the synthetic plastic material."[73] In the same year Charles F. Reeves, vice-president for sales of Celluloid Corporation, lectured his colleagues in "Profits from Depression." By considering "the sales possibilities of color and beauty," businessmen could reap "untold profit" from the Depression. His three examples—cosmetic jars, clock cases, and electrical appliances—had gained consumer acceptance through use of Lumarith, a Celluloid product.[74]

More sophistication appeared in a campaign mounted by Bakelite Corporation to expand its product beyond the electrical industry, which used Bakelite for insulation. In 1932 the company held a symposium to acquaint industrial designers with Bakelite's potential as a material for consumer goods.[75] Soon afterward the corporation began advertising in *Plastics and Molded Products* and *Sales Management* to sell manufacturers on redesigning products with Bakelite, "the material of a thousand uses." A headline in the first ad stated directly, "You can profit from the vogue for simple designs." Modern taste preferred "simplicity of line" to "purposeless ornamentation." More important, such "attractive modern designs" entailed "less costly molds." [76]

Here the Bakelite Corporation addressed a major problem of product design during the Depression. Not only did manufacturers have to attract customers by means of appearance. They also had to reduce production costs to meet the lowered prices consumers were able to pay. As *Product Engineering* phrased it, the Depression required "better products for less money." [77] Plastic met this need in two ways. First, owing to the technological innovation typical of a new industry, its price fell in relation to other materials. Between 1931 and 1935 the volume of coal-tar resin production tripled, while prices fell by 20 percent. [78] Second, according to Franklin E. Brill of General Plastics—a major publicist for design in plastics—manufacturers could avoid the high cost of hand-engraved molds by "using simple machine-cut forms to get that verve and dash which is so expressive of contemporary life." By emphasizing contours rather than details, by sandblasting molds for texture, and by creating contrast through the use of cheap metal inlays instead of expensive engraving, a manufacturer was able to provide modernism at reduced cost. Even plain black Bakelite, far cheaper than colored plastics, acquired beauty through polished curves, ribbing, fluting, and setbacks—all easily machined into a mold. [79] Designer Van Doren once suggested that economic necessity determined the modern style's geometric forms, but it seems likely that they coincided by chance. [80]

Whatever the case, Bakelite invested heavily in industrial design. Its advertising continued in 1933 and 1934 with full-page spreads devoted to products containing Bakelite and their designers. "Raymond Loewy says," ran a typical headline followed by a quotation: "The 'New Deal' in industrial design is establishing the triumph of beauty through simplicity." Before-and-after shots of the Gestetner duplicator accompanied a text intended to suggest the possibility of increased profits through the beauty of Bakelite brought out by an industrial designer. A small photograph of Loewy and a brief sketch of his career created an image of the designer as celebrity. Similar advertisements featured a washing machine by Dreyfuss, a telephone index by Geddes (figure 51), and a Coke dispenser by Vassos (figure 52). While the campaign encouraged manufacturers to use Bakelite as a stylish material for consumer goods, it also boosted the reputation of industrial design and of particular designers. [81]

This design emphasis spread through the plastics industry. The Marblette Corporation proclaimed costume jewelry made from extruded plastic rods, available in three hundred shapes and two hundred colors, to be "leaders of style." General Electric exhorted manufacturers to "plan for profits" by redesigning products with Textolite. And American Cyanamid lauded the design success of the L. H. Hyatt Company's restaurant cream dispenser, which "offered the vivid contrast of gleaming white and jet black" Beetle. [82] Even the industry's trade journal capitulated to the design movement. After an interlude as *Plastic Products* it appeared reborn in September 1934 as *Modern Plastics*, with pages graced by new typography, layouts, and higher graphic standards provided by Joseph Sinel. Soon it acquired a new editor, E. F. Lougee, who virtually eliminated technical articles and emphasized instead industrial design. In 1934 and 1935 Lougee himself profiled Dreyfuss, Guild, Loewy, and Gilbert Rohde. He opened the magazine to designers, spotlighted products made with plastics, and in 1936 began an annual design competition. The new magazine functioned

51. Geddes. Telephone number index for Bates Manufacturing Co., 1933. Geddes Collection.

52. John Vassos. Soda fountain dispenser for Coca-Cola, c. 1934. Crain Communications, Inc.

as a clearinghouse for the plastics industry and consuming manufacturers by stressing increased profit through redesign with plastic. Designers gloried in this generous publicity, so different from that granted in the pages of *Product Engineering*.

Publicity in a wide range of business and trade magazines, even the faint praise of engineers, contributed to an industrial design boom by mid-decade. According to one account, hundreds of manufacturers were "forced" to redesign products both to meet more advanced competitors and to appease a public that required modern style.[83] *Product Engineering*'s annual survey in 1932 revealed that only 48 percent of domestic appliance manufacturers concerned themselves with improving product appearance. By 1935 the figure had risen to 91 percent.[84] Despite attempts to represent design as a rational process, however, some businessmen had an idea that the typical industrial designer was, as designer Peter Müller-Munk recalled, "a wizard of gloss, the man with the airbrush who could take the manufacturer's widget, streamline its housing, add a bit of trim, and move it from twentieth to first place in its field."[85] To some observers the profession seemed filled with "drawing-board four-flushers lured into the field by the hope of cashing in big on empty inspiration," who had no "practical idea of what real design is and what it can and can't do."[86] As the decade entered its second half amidst an economic recovery that made manufacturers less likely to grasp at industrial design as a panacea, designers strove to deflate ballyhoo they had themselves generated. Even *Fortune*'s influential article of 1934 had contained misrepresentations, Van Doren revealed. His public weighing scale (figure 53), which *Fortune* reported had sparked a sales increase of 900 percent in one year, had actually been purchased by two large syndicates replacing older scales leased to commercial outlets—after which the design generated virtually no income.[87] As a result of suspicions aroused by unfulfilled promises, designers aimed publicity even more at convincing manufacturers of industrial design's mundane nature.

They emphasized more strongly collaboration with both engineering and marketing departments. More significant, discussions of beauty became more sophisticated. Many designers had argued that consumers would naturally choose products whose designs approached aesthetic perfection. But now they talked more of manipulating the public through superficial style changes. A designer's job, ran the argument, was never finished. Geddes's assertion of 1932 that good design "adds length of life" to a product "because it takes longer to tire of it" would have seemed naive at best, hypocritical at worst.[88] J. O. Reinecke, a younger designer, warned manufacturers in 1937 that "Design Dates Your Product." He advised them to take out "insurance" against "loss of sales" owing to "the constant changing of tastes" by hiring designers "to meet consumer demands and, to some extent, dictate trends."[89] More blatant, John Vassos told designers assembled at the American Furniture Mart in Chicago that "the world depends on obsolescence and new merchandise." As designer of RCA radio cabinets, he had broad duties that included addressing an annual meeting of local dealers. Each year, he related, "I tell them how terrible our last year's line was . . . because I like to try and make the new line better than last year's."[90] It might be true, as E. F. Lougee claimed, that industrial design had "played an important part in industrial and commercial recovery," but to remain in business, designers would have to convince manufacturers of the efficacy of continual redesign in good times as well as bad.[91]

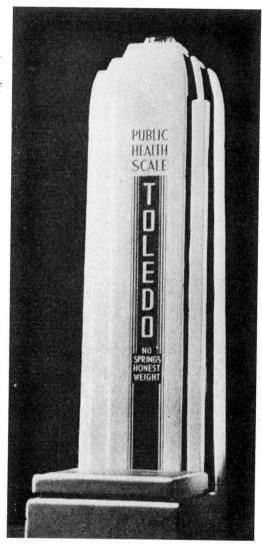

53. Harold Van Doren. Public scale for Toledo Scale Co., 1932. Printers' Ink Monthly.

Industrialized Design

It is not surprising that we all came to approximately the same conclusions, adopted the same methods; because those who did not, simply did not last.
—*Walter Dorwin Teague*

Under the aegis of the Metropolitan Museum of Art, Richard F. Bach held his thirteenth Contemporary American Industrial Art Exposition late in 1934. Although Bach had pioneered the concept of industrial design, he lagged behind in popularizing its fruits. He exhibited mainly interiors with household furnishings designed for the occasion by architects of the New York modernistic school. But one interior differed radically from others. Raymond Loewy and stage designer Lee Simonson collaborated on a mock-up of an industrial designer's office (figure 54). In planning it, they "chose the severest of lines and depended on the materials" alone "for attractiveness" because, as Loewy noted, an industrial designer's office "should be adapted to its function." According to him, it was "really a clinic—a place where products are examined, studied and diagnosed."[1] Their mock-up did resemble a hospital room, with washable walls of ivory Formica, a blue linoleum floor, indirect lighting supplemented by spotlights, and tubular furniture. And Loewy himself, posing nonchalantly for a publicity photo, seemed a skilled technician who with a few pen strokes could restore a corporation's economic health. The stage set provided by Loewy and Simonson imbued the designer with the public glamor of a scientist or technician while satisfying expectations of an aesthete like Bach.

Their mock-up lacked sincerity, however, because it *was* only a set, appropriate for a Hollywood film, expressive of the industrial design mystique, but hardly a typical industrial design office. Despite functional trimmings it would have proved a nonfunctional working environment. Hardly large enough for one person, it neglected in its deification of the individual the space and equipment required by architects, draftsmen, modelers, engineers, researchers, public relations experts, and general office help who staffed a typical design office. Bach might protest that quality suffered as "design for industry" became "industrialized design," but he could

54. *Raymond Loewy and Lee Simonson. Mock-up of an industrial designer's office for the Metropolitan Museum of Art's Contemporary American Industrial Art Exposition, 1934. Raymond Loewy.*

not reverse the trend.[2] Adequate service to a number of manufacturers required design factories characterized by teamwork, division of labor, and routine. Illusion and reality appeared ironically juxtaposed in a contract Geddes signed in 1930 with the Philadelphia Storage Battery Company, manufacturer of Philco radios. Their agreement stipulated that each radio cabinet would carry the "NBG" monogram and the words "This Cabinet Designed by Norman Bel Geddes," but it did not "require the ARTIST to attend personally at conferences or to create and execute designs: it being the intent and purpose of this contract that the ARTIST may employ draftsmen, assistants, artists, engineers and other specialists to confer with the

COMPANY and to execute, create, or develop designs under his direction."[3] Later explaining why employees received no credit for their work, Geddes argued that effective publicity depended on the reputation of a name built up over long years.[4] Whatever the realities of their business, he and other designers recognized the value of individual charisma.

Standard Practice

By the time Geddes's firm issued its office manuals in the forties, he had honed its organizational distinctions to an extreme degree. His thirst for efficiency would have pleased Taylor of the factory management movement. At the core of his office philosophy reigned a "belief that specialized routine results in the formation of correct working habits, reduces errors to a minimum, establishes a standard method of work common to everyone in the office, lessens the strain on individuals, is more expeditious than optional individual methods, and frees higher-salaried persons from detail that can be done by lower-salaried employees." The office contained three departments. The management department handled financial and legal matters—particularly patents. The service department took care of both public relations and negotiations with clients. Essentially a sales department, its euphemistic title enabled members to work directly with top executives of a firm instead of with sales or purchasing agents. Finally, the production department conducted the firm's most obvious work.[5] This department itself contained four divisions: design, responsible for generating ideas and coordinating each project; drafting, which included subsections devoted to drawing, printing, and photography; modeling; and technical, whose members concerned themselves with engineering specifications and passed judgment on the practicality of design ideas.[6] A general manager acted as liaison among these autonomous departments, promoted "enthusiastic team play" among the firm's operatives, and oversaw circulation of a forest of memos, checklists, data sheets, and other highly specialized forms.[7] As a further gesture toward cohesion, a schedule board forty feet square dovetailed assignments of each employee for weeks ahead, dilatory individuals being indicated by red markers.[8] Geddes was "a fanatic for organization," as an associate has recalled, and his efforts produced a turnover of bewildered office managers.[9] But his office's paper maze represented by exaggeration the complexity assumed by most successful industrial design firms.

Geddes himself by 1934 had about thirty employees crammed into an east side brownstone equipped with drafting room, modeling studio, and print shop. The firm maintained a library of books on engineering, architecture, and design and subscribed to more than a hundred technical and trade journals. Perhaps owing to Geddes's love of architecture, most of his draftsmen came from architecture schools. Among his key associates were Worthen Paxton, with degrees in both engineering and architecture from Yale; Roger Nowland, a 1927 MIT graduate in aeronautical engineering; Peter Schladermundt, educated at the Yale School of Architecture and a former employee of Raymond Hood; Garth Huxtable, a 1933 design graduate of the Massachusetts School of Art; and Earl Newsom, a Columbia English Ph.D. with experience in market research, who acted as business manager and solicited new accounts. For special projects—such as an assignment from Walter Chrysler to redesign the front end of the 1934 Airflow— Geddes often hired temporary experts—in this case Carl Otto, who had previously worked in the Art and Color Section of General Motors. Geddes subjected their work and that of routine draftsmen and modelers to close criticism, but he tended to focus personal attention on visionary projects and leave mundane assignments to his staff. He emphasized an openness of approach and independence of vision made possible by his own lack of formal education but surrounded himself all the same with graduates of leading technical and professional schools. His firm was not the romantic one-man effort suggested as typical of industrial design by the Metropolitan's office mock-up.[10]

Offices of the other three major designers did not expand as rapidly as Geddes's, but they grew steadily through the decade while his temporarily declined from 1935 to 1937, perhaps owing to his mercurial temperament and distaste for mundane styling jobs. Dreyfuss consciously limited his clients to about fifteen. He concentrated his sales effort on maintaining a select group of clients and providing them with a wide range of continuing design services rather than on attracting numerous one-shot commissions. Even so, his staff expanded. When *Fortune* reviewed industrial design in 1934, Dreyfuss reported having five staff members. By the end of the decade he had fifteen employees, five of whom were designers. Although he supervised his office's work more closely than some designers, he found himself increasingly drawn into correspondence and negotiations with clients.[11]

As the decade progressed, Teague developed a large office. Between 1934 and 1937 his staff grew from four to twenty-five. Physically, the office expanded from two rooms in a residential apartment house to an entire floor, including private office, consulting room, offices for associates, secretarial and file room, drafting room, and model shop. Architecture graduates of universities like the Massachusetts Institute of Technology, Cornell, and New York University comprised his design staff. By mid-decade Teague had gathered a group of lieutenants, many of whom remained with the firm beyond Teague's death in 1960. John D. Brophy joined as business manager in 1932. Starting as a draftsman in 1935, Robert J. Harper advanced rapidly, became director of design in 1940, and in 1945 became Teague's first partner. W. Dorwin Teague, Jr., who as a college student had prepared drawings for the Marmon 16 automobile, joined the office officially in 1934 and soon became director of product design. Lists of drawings in company files reveal that Teague, Jr., and Harper contributed much work and supervised other draftsmen, while Teague himself contributed no drawings of which record remains. In fact, as his son recalled, they often disagreed on design matters—a source of friction compounded by the fact that Teague, Jr., considered himself a more imaginative and technically aware product designer. Teague himself spent much time traveling to consult with clients, soliciting

new business by correspondence, and acting as a publicist. Even in these roles he required an assistant, Martin Dodge, formerly of the Geddes office, who handled routine correspondence and factory trips after negotiation of a contract. Despite his distance from the actual design process, Teague closely supervised his firm's work. Returning from a vacation, he expressed anger that certain cabinet designs presented by Dodge and his son to the Crosley Radio Corporation had not retained "the fine and simple form in which they were conceived."[12] Despite his firm control, his office with its rationalized division of labor did not reflect the elegant simplicity of Loewy's prototype.[13]

Actually, Loewy may have contemplated such an office when he and Simonson collaborated on its design. In 1934 he reported to *Fortune* only one staff member, possibly himself. This limitation derived from necessity. Before designing the successful Sears Coldspot refrigerator of 1935, Loewy had attracted few clients. After this turning point, which coincided with an economic "boomlet," he found his office prospering and created three executive positions, hired a business manager, set up a public relations department, sought permanent technical advisers, and established a model shop. Like Teague and Geddes, and to a lesser extent Dreyfuss, he "began to see possibilities of big business."[14] By 1936 his staff in New York numbered twelve, and he had established branches in Chicago and London. In the same year he hired William Snaith, an architect, stage designer, and interior decorator, who steered the firm into commercial architecture, including store fronts and interiors. At decade's end his office covered a penthouse on Fifth Avenue and an entire floor below filled with sixty employees. As Loewy later recalled, he "always believed that teamwork was absolutely necessary" and enjoyed delegating authority to "young men" who soon became "full-fledged designers, businessmen, getting knowledge of administration, sales." But Loewy himself discovered that "running a large organization" was "not especially pleasant."[15] More than other designers, he retreated to a private life of travel and recreation, leaving business details to associates and concerning himself only with personally attractive commissions, usually vehicles.

Whatever his personal emphasis, no designer succeeded without an extended staff of design executives, technical authorities, draftsmen, modelers, public relations experts, and general office help. Individual idiosyncrasies like Geddes's mania for organization gave each office its own character, but most designers adopted similar design methods. The standard approach appeared clearly in one of Dreyfuss's first mass production assignments, a redesign of the Big Ben alarm clock for the Western Clock Company (Westclox), sometime between 1930 and 1932 (figure 55). After negotiating a contract, Dreyfuss visited the factory in La Salle, Illinois, and studied the firm's manufacturing techniques to discover what design changes might be practical. Returning to New York with several Big Ben clocks, he purchased competitive models and tested them all on his own nightstand to determine the factors that made a clock face easy to read and its mechanism easy to operate. As a rough exercise in market research, Dreyfuss observed customers shopping for alarm clocks in department stores and noted that they usually chose the heaviest, presumably because it seemed most durable.

He and his assistants then worked for days, producing more than a hundred rough sketches of proposed changes in the Big Ben. Aided by his wife, Doris Marks, who continued to play an important role in the firm, he chose the best of these sketches, had renderings made, and took them to the factory for a conference with Westclox executives. Present were not only top officers of the firm but also the production manager,

55. *Four generations of Westclox's Big Ben alarm clock, left to right. Dreyfuss's Big Ben of c. 1931 is second from the right. On the far right is a later Dreyfuss adjustment. Dreyfuss Archive.*

who would have to produce the design; the sales manager, who would have to sell it to dealers; and a representative of the firm's advertising agency, who would have to sell it to consumers. Together they settled on two or three possibilities, which Dreyfuss then worked into three-dimensional models, first of clay, which permitted easy adjustment of proportions and details, and then of wood. Armed, as he thought, with the best of these models and sketches based on it, he traveled again to the factory to present the company with his final design. But his secretary forgot to put the model in his briefcase. Whatever the company's response to this lapse in rational planning, it accepted his work, produced its own working drawings, and tooled up for the new Big Ben. The result was a slightly smaller clock with more refined, less tinny lines, a face brought closer to the glass to eliminate distracting shadows, stylized numerals easier to read than the previous ornate ones, and a plug of iron in the base to make the clock heavier. Dreyfuss had done most of the work by himself, aided by his wife and a few assistants, but an increased volume of more complicated jobs would soon require division of labor. In numerous forums leading designers outlined essentially the same design process.[16]

The first requirement was a client. In the thirties advertising agencies continued to bring together manufacturers and designers. The National Radiator Corporation asked Teague to redesign its line of gas boilers after receiving his name from the Ketchum, MacLeod and Grove Agency of Pittsburgh.[17] Recommendations from satisfied clients contributed equally as much business. The Louis Allis Company requested aid from Teague in designing casings of its electric motors after hearing from the Square D Company of the commercial success of a circuit-breaker for which Teague had provided a housing.[18] Personal connections helped as well. Continuing to profit from his marital tie to the J. Walter Thompson Agency, Geddes in 1936 received a commission for a model of "the

city of tomorrow" to be used in a Shell Oil advertising campaign organized around traffic control.[19] And Loewy began a profitable association with the Pennsylvania Railroad on the strength of a letter written to its president in 1934 by mutual friends Charles and Stuart Symington.[20] Most designers also relied on direct solicitation to drum up business. Loewy obtained his crucial Coldspot refrigerator job on the strength of having "called on and impressed" a Sears executive.[21]

Most important, designers relied for future business on more commissions from satisfied clients. After designing the interior of Ford's exhibit at the Century of Progress Exposition, Teague pressed Edsel Ford to let him apply his "fresh and detached point of view" to the design of auto bodies. Ford declined, but, owing to his "great satisfaction" with the exhibit, he continued to employ Teague's firm for major expositions, auto shows, permanent showrooms and offices, and even his own home.[22] Teague enjoyed similar long-term arrangements with Eastman Kodak, the National Cash Register Company, and the A. B. Dick Company, manufacturer of office machines. Other designers profited from permanent relationships. Dreyfuss had retainer agreements with Bell Telephone Laboratories, the Crane Company (bathroom fixtures), General Time, the Hoover Company, John Deere, *McCall's*, and the New York Central Railroad; while Loewy counted for continuing employment on Greyhound, International Harvester, the Pennsylvania Railroad, and Studebaker. It was a measure of Geddes's instability that he attracted no permanent clients.

Most contracts resulted from a direct meeting of a designer with the president of a client corporation. Geddes, in fact, refused to meet initially with lesser figures.[23] No case has come to light of a designer rejecting a job because he did not feel adequate, but Dreyfuss advocated careful consideration of a manufacturer's problem to ensure that his office could contribute.[24] More often, a designer used all the arts of persuasion he could muster to convince a manufacturer that, as Teague told the general manager of the National Radiator Corporation, redesign provided "an opportunity to promote your business by developing the eye appeal of your products."[25]

After a successful meeting of minds, the two parties drew up a formal contract, stipulating services to be provided by the designer and fees to be paid by the manufacturer. Smaller jobs usually cost a flat fee. For a gas boiler Teague received $1,000 plus travel expenses.[26] Geddes charged the same for redesigning a telephone number index (figure 51). Included in his preliminary estimate were labor charges of $225 for sketching and modeling (at $15 per man-day), $130 for Paxton's supervision (at $65 per day), and $40 for a consolidated day of his own time; an overhead charge of $395 (equal to total labor cost); and $200, or 25 percent of total cost, as his office's profit.[27] For an entire line of goods, contracts sometimes specified payment on a cost-plus basis but more often embodied a monthly retainer —which covered a designer's expertise and a margin of profit—supplemented by actual costs of labor, materials, overhead, and transportation. Geddes had such a contract with Philco. Intended to last five years, it provided for an annual retainer of $25,000 plus actual costs (materials, salaries, a proportion of overhead, and travel outside New York), not to exceed $3,000 a month for four months and $1,500 a month the rest of the year (for a possible maximum of $49,000 a year). The contract provided royalties of a few cents per cabinet beyond a substantial number of sales, but this practice was not widespread. The company terminated the contract at the end of a year after Geddes had produced fourteen designs at a maximum cost (excluding any royalties) of $3,500 apiece.[28] As the profession settled down, most designers charged modest fees to counter adverse rumors of six-figure sums. When Teague

solicited business from General Electric in 1936, he received a cool response. The corporation had used outsiders in the past but found that "prices had been exorbitant."[29]

Signing a contract usually led to formation of a core group of two or three designers, depending on the size of the project. For a period of a week to several months they studied the product to be redesigned—its appearance, mechanical function, and marketing and advertising techniques for it and its competition. They dismantled a sample of the current model to determine if its undesirable appearance derived from its mechanical arrangement. If functional parts yielded poor appearance, they sought new arrangements of parts that would facilitate redesign and be cheaper to produce and assemble. The team also obtained and analyzed competing products or, if expense prohibited, examined photographs. From industry-wide sales figures, the team deduced developing trends that might be expected to last. Advertisements and sales brochures of all makes revealed common sales points and suggested neglected factors that might be emphasized to project an aura of novelty around a new design. Dreyfuss, as noted previously, personally observed customers at the point of purchase to determine factors commonly considered. Other designers conducted formal market research, most often with questionnaires. After total immersion in a client's business, a design team toured the plant.

A factory tour served two purposes. First, it enabled a design team to assess a client's manufacturing capabilities and determine what changes in forms, materials, and finishes could be accomplished without prohibitive expense. More important, it provided designers an opportunity to instill confidence in people they would work with during the design process—research engineers, product engineers, production-line engineers, sales managers, and advertising directors— any of whom could jeopardize successful association. Sometimes designers never established rapport. Socony (now Mobil) never used a standardized prefabricated service-station design developed by Geddes because too many insiders feared "the wiping out of their own design and construction department."[30] Not only product engineers proved recalcitrant. Loewy found that sales managers typically underestimated public taste and often "insisted on making an item look cheap and sleazy."[31] Among other things, a factory tour let designers know whom they could trust and who might later cause trouble. Often, conversion of a few key individuals ensured smooth sailing.

After assimilating data gathered in the office and at a client's plant, a design team devised several solutions to the problem and set draftsmen to churning out rough sketches. The best of these received careful treatment as renderings and were submitted to the client at a meeting of all department heads. Usually they settled on two or three for further development. At this point, attention focused increasingly on technical rather than visual details—materials, manufacturing methods, and appropriate finishes. Often designs changed considerably in a continuing give-and-take process between design team and client. Concurrently, modelers developed the exact form of the final design, working first in clay. When the head of the office, if he was not a member of the team, had given approval, a full-scale model of wood or plaster with meticulous measurements was presented to the client for inspection. Sometimes a designer then prepared working drawings, blueprints, and a working model from which dies and molds could be prepared. Just as often the client handled these final details. Often a designer provided such extras as a new trademark or logo, packaging, retail sales displays, and assistance in preparation of sales literature, advertising layouts, and promotion. Throughout these steps teamwork and collaboration remained essential. This plurality contributed as much as intention to a general impersonality of style. It also won industrial

design a reputation as a legitimate, practical profession capable of serving business.

Without careful analysis of an exhaustive number of cases, one could not weigh the success of these procedures. Most small companies were not equipped for complete exploration of market trends, fitness of a product in terms of salability and operability, use of new materials like plastic and lightweight alloys, which often lowered production and shipping costs, and use of unfamiliar processes or materials common in other industries. Even large corporations with their own design staffs often benefited from a fresh point of view. Recently finished with the Hupmobile, for which he had specified an aluminum radiator grill, Loewy provided the Sears Coldspot refrigerator of 1935 with aluminum storage racks—cheaper to produce and more functional because they would not rust.[32] And Dreyfuss, designing a tractor for John Deere, cut costs by eliminating the customary hood after observing that most farmers discarded it when they made the first repair.[33] Such individual touches seem minor in the context of an entire job, but they demonstrated expertise and boosted the reputation of the design profession. The best evidence of their method's success was the fact that they did prosper during the Depression. In general, manufacturers did appreciate their work and, in many cases, they provided further commissions—an ultimate test of satisfaction.

But commercial success often came at the cost of compromising the functionalist design philosophy espoused by most designers. Geddes admitted as much when he stated in his office manual that the first steps of any commission included determining whether a client was "more interested in functional design or visual appearance."[34] Most designers occasionally hit each extreme but usually struck a balance, depending on the job at hand. Teague tipped toward appearance with the WIZ Autographic sales receipt register, redesigned for the American Sales Book Company in 1934

56. WIZ Autographic sales receipt register,
American Sales Book Co., before redesign. Pencil Points.

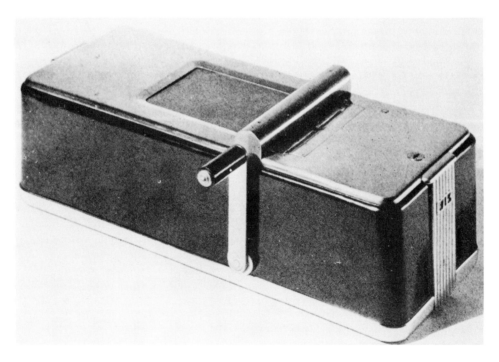

57. WIZ Autographic sales receipt register, after
redesign by Teague, 1934. Pencil Points.

(figures 56 and 57). Planned to be attractive enough for use in fashionable shops, his case design necessitated "a complete redesign of the working mechanism" by engineering consultants hired by the company—clearly an instance of jamming a machine into an ill-fitting housing.[35]

In another instance his firm swung all the way to the functional side. In 1936 the Louis Allis Company approached Teague for a uniform and "pleasing appearance" in a line of electric motors. A sales manager suggested that "when we get a design of a casing that looks good to us, we will then try to put the necessary insides into this casing." Teague balked, wanting instead "to develop our design around these necessary insides."[36] Eventually his office became involved not only with standardization of exterior housings of various types of motors but also, as a cost-saving measure, with standardization of interior functional parts throughout the line—necessitating development of new cooling systems. Teague never completed any designs because the variety of motors proved too diverse for standardization, but his office did investigate complex aspects of electrical engineering. Rarely did designers become so technically involved. Dreyfuss exemplified "standard practice" when he noted that his office often suggested new engineering ideas but left clients to develop them. Most patents issued to designers granted rights to external forms rather than actual inventions and received separate numbers as "design patents." Although designers talked and wrote about function, they dealt primarily in appearance.

Teague's Platonic design philosophy received hard knocks in compromises his office reached with clients. His repeated assertion that industrial design involved discovering the one right form latent within a given type of product or machine contradicted the major reason given by manufacturers for employing his services—stimulation of sales through creation of artificial obsolescence. Eventually the best-conceived design would have to yield to a successor if only for novelty's sake. In 1936 the Taylor Instrument Company replaced his Stormoguide barometer of 1931, widely heralded by design publicists as the epitome of restrained modernism, with a model developed by the company's own engineer and boosted as having "an appearance of life and sparkle previously lacking" (see figure 58).[37] But the irony of this case dwindled in comparison to his experience with the American Gas Machine Company.

Traditionally a manufacturer of camp stoves and lanterns, American Gas Machine had sought year-round sales in 1932 with an oil-burning space heater for the home. After an initial season of dismal sales, the company hired Teague on the recommendation of the N. W. Ayer Agency. Most manufacturers patterned space heaters after radio cabinets to make them blend into living rooms, but their tinny sheet-metal and imitation-wood finishes looked cheap. Teague decided that a heater "should look like a heater."[38] And since no one knew exactly how a heater should look, he set the standard (figure 59). Although he retained the old model's boxy shape, he camouflaged obtrusive sheet-metal seams by wrapping the two side panels several inches around the front and running chrome strips down these joints. Thus the heater seemed constructed of a continuous piece of metal. Instead of disguising its doors as a Gothic-inspired speaker opening, he set them in a chrome frame and recessed them to give the heater an appearance of thickness and solidity unusual for sheet metal. A dark brown lacquer replaced the usual imitation-wood finish. This design, introduced in the fall of 1933, triggered an increase in annual sales of 400 percent, gained the company distribution in fashionable furniture stores, and, as the company later claimed, set the pace for an entire industry. Teague had found the perfect design.[39]

But three years later American Gas Machine requested a new design from Teague. As if aware of the earlier heater's

58. Teague's Stormoguide barometer (left), 1931,
and its replacement (right), designed by Taylor
Instrument Co.'s own engineer in 1936.
Reprinted by permission of Modern Plastics
Magazine, McGraw-Hill, Inc.

59. Teague. Oil-burning space heater for
American Gas Machine Co., 1933. Crain
Communications, Inc.

definitive nature, company executive Russell N. Hanson explained that it was "not a matter of our being dissatisfied with the present design as everyone in our organization, as well as the dealers we have contacted, is enthusiastic about it." But they did "feel the necessity for some sort of change in models each year in order to maintain the enthusiasm we have been able to maintain for our line of heaters."[40] After Teague accepted his new commission, Hanson sympathized with his plight in having to tinker with a perfect design. "Personally," he wrote, "I don't know just what you are going to do to improve the appearance of this year's line, but that is your headache, not mine." In conclusion he confused matters by suggesting that retooling costs could be lessened by retaining the current door and frame, unless "you think of something better."[41] Teague did design the 1937 model, specifying a round door to replace the old rectangular one.

Several months later Hanson went through the same routine with a potential 1938 model. Noting that the 1937 heater had involved an investment of $6,000 in dies for the top grill, he suggested using the same part again—but only if it "works into your new design." In fact, Hanson emphasized, there was "not the slightest insistence on our part that you maintain the grill as it is." And then, as if he had read Teague's essays on discovering the ultimate form of a product, he apologized again. "I almost hesitate to mention it," he wrote, "because I want the new heater to be your conception of what it should be." As an afterthought Hanson timidly broached the possibility of retaining the round door of the previous year's model.[42]

In his design for 1938 Teague did retain these elements. Later he learned that the company had not produced it for the astonishing reason that it "was too much like the 1937." The company wanted a completely new design for 1939. Even more, it wanted a hypocritical statement from Teague—to be used by salesmen—that the

old 1937 model remained unchanged in 1938 because it was "so soundly designed" and "so inherently 'right'" that it remained superior to new models introduced by competitors.[43] Providing more than requested, Teague included a photograph of the 1937 space heater and a discussion of his work for the company in an article about his Platonic design philosophy. In any application of industrial design, he insisted, there was only one "right solution to the problem—not a dozen solutions, but one right one."[44] But in practice he provided as many annual models as a client requested. His failure to satisfy American Gas Machine derived not from refusal to compromise his stated design principles but from the company's desire to minimize retooling costs by retaining elements of previous models. Teague must have been shocked when Hanson announced that the company was also rejecting his design for the 1939 model—because "there was too much similarity between it and the design which we have marketed for the past two years." With scores of manufacturers "diving into the heater game," the company had to depart from an industry-wide design trend that Teague himself had set by introducing a model "distinctively different from the products of our competitors."[45]

The extreme ironies of this case demonstrated the compromises necessary in a profession that espoused idealistic aesthetic standards but existed for the sole purpose of serving businessmen who were concerned chiefly with profits. No doubt Teague sincerely believed in his ideals—particularly when he drew from them an abstract vision of a machine-age world transformed into a place of efficient, harmonious living—but he had to follow the "standard practice" of his profession. The industrial designer, as Dreyfuss once insisted, was indeed an artist, but he had deserted the "ivory tower . . . above the multitude." He had "taken [the] elevator to [the] ground floor."[46] After all, industrial design was itself a business, and in business the customer is always right.

Stealing All the Credit

Despite laudatory case studies published by magazines like *Sales Management* and *Modern Plastics*, it remains difficult to assess industrial design's contribution to economic recovery. To show that industrial design had a positive effect on the economy, one must prove that consumers bought more products in general than they would have if redesign had not become standard among manufacturers. Since this is impossible to prove, one might see if redesign improved the sales of specific products. Is it significant that Loewy's 1935 Coldspot refrigerator or Geddes's Bates telephone number index enjoyed vastly greater sales than immediate predecessors? Did they stimulate consumption in a real sense that might be reflected in the gross national product? Or did they simply lead consumers to allocate for their purchase funds that would have been spent regardless? Looking at the whole economic picture, one must conclude that industrial design's economic significance cannot be proved beyond such an internal bookkeeping effect.

On the other hand, if it can be proved that industrial design enabled some manufacturers to obtain larger shares of a fixed amount at the expense of others, then industrial design is revealed as an important cultural force even if its economic significance was limited to the relative fortunes of individual firms. If consumers did choose the 1935 Coldspot over a boxier refrigerator, then one can conclude that its appearance— its style—reflected some image with which people chose to identify. And since this image or set of images derived from gleaming streamlined vehicles, symbols of technological progress, then one can conclude tentatively that industrial design did provide the dismal economic scene with a suggestion of optimism. But first one must prove that industrial design was responsible for increases in sales of individual products.

Marketing a product involved so many factors that contemporary observers had difficulty assessing the significance of redesign. In fact, Van Doren, accusing his fellow designers of having a "habit of stealing all the credit" for a successful product, declared that it was impossible to determine a designer's role in sales. Even the case of the Loewy Coldspot, which had apparently brought Sears from tenth to fourth in the refrigerator industry, was unclear. Van Doren admitted the importance of the design itself, but he cited more efficient promotion, a larger advertising budget, and a company goal of boosting its refrigerators out of the red as equally important factors.[47] A study published in 1942 concluded that advertising's effect on sales could not be measured—for reasons that apply as well to industrial design. According to this study, a product's sales record was affected by changes in buying power, product design, price structure, effectiveness of sales force, promotional methods, advertising, packaging, number of competitors, fashions, distribution methods, and living habits of the public. There was no way to assess the relative importance of these variables in a given case; additionally, in order to reach a true evaluation of the situation, one would have to conduct a similar analysis of all competing products.[48] Still, the argument for design often seemed conclusive, if only by intuition. President W. Frank Roberts of the Standard Gas Equipment Corporation concluded that although there had been an "improvement in general business conditions" since the introduction of a stove designed by Geddes, it was "also absolutely a fact that our increasing sales were very largely due to the new range."[49]

Another complication arose from the fact that domestic appliances provided the most spectacular industrial design success stories. Unlike the automobile market, whose saturation contributed to the Depression, the electrical appliance market continued to expand. The example of the mechanical refrigerator is revealing. Not until 1930 did the number of refrigerators produced annually exceed the number of traditional iceboxes. In that year only a tenth of the 68 percent of American homes wired for electricity were equipped with refrigerators. Consumption of refrigerators could only go up. Annual sales dipped slightly in 1932 and drastically in 1938 in response to general economic conditions, but for the most part sales increased steadily through the decade. Annual sales rose from 791,000 in 1930 to a high of 2,310,000 in 1937 (not topped until 1940). The average price of a refrigerator fell from $275 in 1930 to $170 in 1933, a figure that remained stable until 1940. The greatest decrease came in 1932, when the average retail price of a refrigerator dropped from $258 to $195 as manufacturers drastically cut prices to increase demand in the worst year of the Depression.[50] Economic pressures, technological innovations, and high-volume production combined to force prices down to a level that made it possible for more people to purchase refrigerators. Despite the Depression, availability of refrigerators and other appliances made the thirties, in Siegfried Giedion's phrase, "the time of the democratization of comfort."[51] In other words, the success of any specific redesigned appliance had to be read against a backdrop of expansion.

Often a designer took credit for a successful product when in fact it enjoyed increased sales because the manufacturer had lowered its price. In a lead article published in 1931, *Product Engineering* noted that the product engineer's main responsibility during the Depression was to reduce production costs in order to meet the low prices that hard-pressed consumers were willing to pay. Since the "efficiency engineer" of the factory management movement had already minimized shop labor costs, savings had to be sought in less expensive materials and assembly methods. *Product Engineering* described this approach as manufacturing "better products for less money."[52] A later historian concluded to the contrary that

"manufacturers very often cheapened the quality of their product while maintaining an illusion of brand consistency and value."[53] However one defined the process, a great many redesign jobs involved reducing costs rather than improving appearance.

The radio industry was the best example. As a purveyor of the new American tempo, the radio enjoyed continually climbing sales figures in the twenties. After the market crash, manufacturers slashed prices from an average of $133 in 1929 to $87 in 1930 in order to reduce inventories. In 1930 several small companies introduced varieties of the "midget" radio or "jalopy" at about $60. These small table models cost less to produce because their compressed circuits and subassemblies required less metal and solder. They also had pressed-metal cases that did not require the "ancient and honorable [but expensive] arts . . . of cabinet making, wood finishing, painting, and lacquering."[54] By 1933 the average cost of a radio fell to $35 after introduction of the even smaller "baby" or "peewee" radio, often as small as $8 \times 6 \times 4''$. Usually priced at about $10, it soon captured half the market in volume of sales. Through such economies of size and material, manufacturers could lower prices enough to make radios available, even during the Depression, to most families.[55]

Many companies hired designers to create cabinets for the smaller, less expensive models. Van Doren and Rideout produced the first plastic radio cabinet for the Air-King Products Company (figure 60). Available in Plaskon of various colors, it looked stylish, with its sharply stepped sides, vertical ribbing up the center, and horizontally ribbed base. Van Doren and Rideout may have won Air-King some customers who would have bought another brand, but they probably did not convince anyone to purchase a radio who was not already planning to. Air-King hired Van Doren in the first place because his skill in working with plastic was well known. And Air-King and other small radio companies turned to plastic not for its polished,

60. Harold Van Doren and John Gordon Rideout. Plastic radio cabinet for Air-King Products Co., 1933. Reprinted by permission of Modern Plastics Magazine, McGraw-Hill, Inc.

machined beauty but "to reduce production costs, and to eliminate finishing and assembly operations."[56] The phenomenal increase in sales derived not from design but from radically lower prices. Design helped only by distinguishing one model from another.

In the final analysis industrial design served as insurance. A product that had remained unchanged since 1920 suffered at the sales counter because of its outmoded appearance. If no company had ever modernized its products, then industry would not have needed industrial design. But once style, and a concept of *changing* styles, became associated with products formerly noteworthy for utility, then industrial design became inevitable. In 1931 American business supported 5,500 product designers, draftsmen, developmental engineers, and researchers. The following year this figure fell to 3,800, but it then rose steadily for the next four years. In 1936 there were 9,500 people directly involved in product development.[57] By then even mail-order houses, bastions of conservative merchandising, had embraced redesign. In 1934 Montgomery Ward had hired Anne Swainson, formerly with Revere Copper and Brass, to head a bureau of design. Under her direction designers created unified logotypes and packages to replace those of the firm's myriad suppliers. Swainson evaluated products sold by the company, produced new designs to replace those found insufficient, and presented suppliers with the new designs. Results appeared in 1936 in a new catalogue, representing an investment of $4 million, with photographs of products instead of traditional woodcuts. The lowest common denominator of American merchandising had embraced industrial design, probably for negative reasons, as a form of insurance against competitors.[58]

Thus industrial design became "standard practice" among manufacturers as they kept up with the Joneses among them. Not a panacea, as many had hoped, it had become a necessity, almost an addiction, withdrawal from which meant loss of sales. To maintain the same effect, users required increased dosage in the form of more frequent re-redesign. Consumers themselves, accomplices as well as victims in this trafficking, suffered an acceleration of artificial obsolescence in part brought on by their demand for novelty. As a result, the appearance of consumer goods oscillated toward a forever-receding equilibrium. Just as the public registered through purchases its satisfaction with a certain type of product, manufacturers engineered further consumption by taking the product or its appearance to the next logical step. Although manufacturers did apply artificial stimuli to elicit consumption responses, they also—at least those rewarded with success—directed those stimuli toward perceived consumer predispositions. Although consumers faced considerable manipulation, the style defined by their choices reflected their outlook on life. In a sense manufacturers and designers went along for the ride when they adopted streamlining.

Everything from a Match to a City

... never before in the history of the world ... has the visual environment undergone so great a willed change.

—Sheldon and Martha Candler Cheney

Mass-produced consumer goods remained central to industrial design, but most designers branched into other areas. Their work generated trademarks, sales brochures, retail displays, and packaging. But their other activities bore no direct relation to product design. Transportation companies—railroads, bus lines, steamship lines, and even the new airlines—sought to attract travelers with comfortable modern interiors and exteriors expressive of machine-age speed. Railroad design in particular extended beyond superficial application of paint to the styling of locomotive and car shapes. Airplanes and ships, whose forms depended vitally on function, did not lend themselves to such treatment.

Other purveyors of services rather than products drew on designers' expertise. Restaurants and gas stations benefited from their experience. Manufacturers of X-ray machines and dental chairs employed industrial design to make their products more attractive and comfortable—not to appeal to purchasers but to provide patients with intangible satisfactions. Department stores and other retailers hired designers to provide exterior shells and interior environments conducive to sales. Corporations called on industrial designers for offices, conference rooms, and display rooms that would facilitate selling to distributors. Finally, designers also provided companies with trade-show displays and exposition buildings calculated to win public confidence under a guise of public-service information. As Teague explained in a business journal, the profession dedicated itself to easing "all sorts of points-of-contact between manufacturer and public where the public is influenced by what it sees and feels."[1]

By 1939 supplementary activities made up a significant proportion of industrial designers' work. Loewy estimated that in the late thirties product design made up 40 percent of his office's work. Other activities included vehicles (30 percent), packaging (20 percent), and interior design of offices and stores

(10 percent).[2] But without the phenomenal success of their product designs—especially domestic appliances—early in the decade, designers would never have branched into other fields.

Cleanlining

Fortune devoted space in its review of industrial design to a kitchen stove designed by Geddes for the Standard Gas Equipment Corporation (figure 61)—which served as prototype for other designers' commissions.[3] Late in 1930 the company offered Geddes

$1,500 for rough sketches of potential designs. He demurred, requesting instead a retainer of $25,000 plus costs and royalties for developing a new design in depth. A survey of 1,200 housewives and domestics revealed ease of cleaning as their main consideration in judging stoves. Geddes therefore gave his stove a skirt that extended to the floor, panels for covering burners and controls when not in use, and, for psychological appeal, a finish of white, "the most sanitary color." This use of white set a trend in an industry that usually offered either the dingy black of cast iron or marbleized sheet metal.

Geddes suggested two innovations for

61. Geddes. Sheet-metal stove for Standard Gas Equipment Corporation, 1933. Crain Communications, Inc.

reducing production costs. Few stove manufacturers had switched from cast iron to sheet metal because its enamel finish often cracked in shipment. Geddes solved the problem by replacing rigid bolted joints with a technique borrowed from skyscraper construction—hooking stove panels to a tubular frame flexible enough to absorb transportation stresses. He also proposed a system of modular construction. Standardized components—broiler, oven, storage compartments, burner assembly, and so on—would be scrambled to create quite different stoves. Geddes gained publicity for this radical innovation, but the company found it cheaper to manufacture "a single-piece stamped steel front frame" for each model.[4]

A novel promotional campaign devised by Geddes became common among industrial designers for several years. As with his Philco radios, each stove bore his monogram and a statement that he designed it. In a sense the stove was signed. A five-piece mail campaign introduced the new stove to dealers. The first mailing praised Geddes as a designer and illustrated earlier work. This brochure described Geddes variously as an "architecturalist," a "designer," and an "artist industrialist." Not until the second mailing did the stove receive more than passing reference. When it reached showrooms in 1933, advertisements mentioned its convenience and ease of cleaning, but they headlined its beauty. The stove heralded "a complete New Deal for the forgotten kitchen." Another ad described it as "a new and beautiful gas range conception created by Norman Bel Geddes." A third announced, "We asked an artist to design the most beautiful, the most practical Gas Range in all the world—*and here it is!*"[5] Geddes's name sold the stove to the public, and the stove sold him to other clients. Other designers benefited, too, as appliance manufacturers realized that a "name designer" provided a selling point for consumers in search of "class" associations. After all, as Standard Gas's advertising manager asserted, there had not "been five

ranges manufactured in the last twenty years that have received as much free advertising combined as this Geddes model."[6]

Another "signed" appliance was the Sears Roebuck Toperator washer (figure 62), designed by Dreyfuss in 1932 and introduced the following year with much fanfare. His name alone constituted "a significant merchandising factor."[7] Sears had reason for associating its product with images of refinement. Still fighting a reputation as a farmers' mail-order house, the company was also engaged in a washing-machine price war. By employing Dreyfuss, Sears hoped to obtain a distinctive appliance that could be priced up and out of intense competition. This ploy became common among manufacturers who employed consultant designers. If a larger manufacturer maintained models in several price ranges, a single high-priced designer-created model lent prestige to cheaper models. For smaller manufacturers, fearful of bankruptcy, concentration on expensive designer models often reduced competition to a minimum.

According to *Forbes*, Sears sold 20,000 Toperators in six months and precipitated a "stampede" for new designs among other manufacturers. *Sales Management* reported the company unable to keep floor models in stock. Obviously a success, the washer exhibited Dreyfuss's careful blending of functional and psychological considerations. Unlike other brands with controls placed awkwardly, his design concentrated them on the wringer arm on top—hence the name Toperator. Most washers had two visible exterior parts, a tub with a motor housing below. The crack between collected moisture, dirt, and mold. Addressing the current clinical emphasis on cleanliness, Dreyfuss enclosed tub and motor in a single shell of mottled blue-green enamel. To cover dirt-collecting bolts, he circled the shell with three polished metal bands. The inside of the tub remained white, as Dreyfuss noted, because "the place where the clothes were to be washed should look as clean as possible."[8]

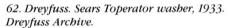

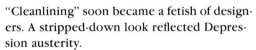

*62. Dreyfuss. Sears Toperator washer, 1933.
Dreyfuss Archive.*

*63. General Electric's monitor-top refrigerator,
1932. Arens Collection.*

"Cleanlining" soon became a fetish of designers. A stripped-down look reflected Depression austerity.

The design history of the General Electric refrigerator also revealed an attention to clean-ness, both in reality and in appearance.[9] As introduced in 1926 and updated in 1932 (figure 63), the appliance resembled its predecessor, the icebox. Heavy, boxy, and squat, it had hardware adequate for Fort Knox. On top sat the condenser's round grillwork, known as a "monitor top" because it resembled a Civil War gunboat. A more functional-looking object could not be imagined. The heavy box indicated insulation, while the condenser witnessed visually that the unit was not an icebox but a mechanical refrigerator. This honestly expressive design fell victim to an inexorable trend toward smooth lines and elimination of visual details—even functional parts.

About 1930 General Electric hired Geddes to assist its own staff in designing a new refrigerator. Always a few years ahead of trends, he took "a very positive, firm stand" against the monitor because it was hard to clean.[10] But the company refused to eliminate the monitor—a trademark—and Geddes contented himself with modifying its appearance. Most of his preliminary sketches revealed the same squat box, lighter hardware, and a revamped condenser resembling one of Eric Mendelsohn's sweeping department-store facades. But General Electric rejected Geddes's suggestions and produced instead a 1932 model that differed little from its original 1926 refrigerator.

In 1933 GE capitulated to the trend of eliminating fussy details. Henry Dreyfuss, replacing his mentor Geddes as consultant, radically redesigned a 1934 model (figure 64) by hiding the condenser in the unit's

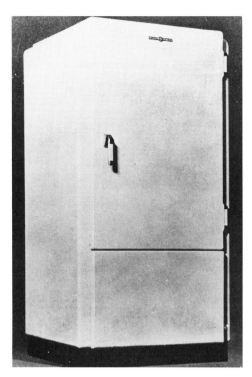

64. Dreyfuss. Refrigerator for General Electric, 1934. Dreyfuss Archive.

base, extending sides to the floor, and providing unobtrusive hardware. The new refrigerator, Dreyfuss informed the nation over network radio, was conceived with "a definite idea in mind: to allow the utility of the machine and the materials of which it was to be constructed to dictate the appearance." But by discarding the monitor, Dreyfuss eliminated the most obvious evidence of utility. Its lines now derived not from function but from a desire to make the unit visually attractive. Its "general proportions," Dreyfuss explained, "were inspired as a piece of sculpture is inspired."[11] Simplification of exterior lines reflected simplification or streamlining of domestic tasks like cleaning. Beauty—and function—were in the eye of the beholder.

Advertised as "styled for the years," the new GE refrigerator remained basically unchanged until 1939. Although GE had learned to "cleanline" products, it had not yet learned to promote artificial obsolescence through annual model changes. Sears Roebuck was not so backward. Its series of Coldspot refrigerators, designed by Raymond Loewy, demonstrated acute awareness of the advantages of continual restyling. In 1932 Sears asked Loewy to design a replacement for a boxy old model that stood on flimsy-looking two-sided stamped legs and had moldings interrupting vertical continuity of line. Loewy gave his new design (figures 1 and 65) an appearance of stability by widening and lowering its legs. Horizontal lines between the door and its frame remained, but Loewy emphasized vertical continuity with slender hinges and three parallel ribs running from top to bottom. When its door was open, chrome ribbing seen on freezer compartment and storage drawers reinforced exterior verticality. The new Coldspot, introduced in 1935, made Sears a leader in the industry. Unlike GE executives, however, those at Sears did not "rest for a time on their laurels."[12] They kept Loewy busy making minor changes in his designs for models introduced in the following three years and raised his annual commission substantially.

A "case history" written later in Loewy's office rationalized the continued redesign of the Coldspot. Sears executives "might have been dubious about the possibilities of a new and better looking box" because Loewy had presumably designed "a 'perfect' refrigerator." But the designer himself did not see his design "as a masterpiece, but as a step in the evolution toward perfection." Therefore he advised "scrapping" his "commercial triumph" for "a new design that would have no connection with the previous one in reference to design treatment."[13] Actually, Loewy's 1936 model (figure 66) differed slightly from that of 1935. The ribbing disappeared, replaced by a single chrome strip down the front. A round nameplate on the door, intended to rectify a design defect by protecting the finish at the spot where one would push the door shut, replaced a rec-

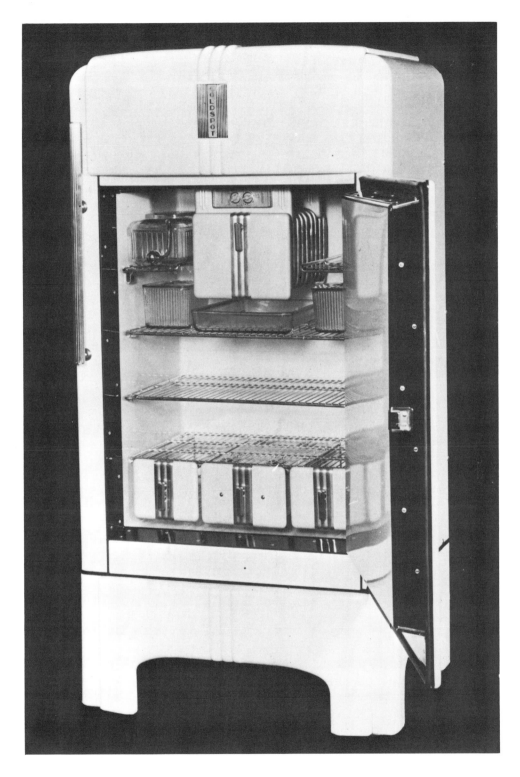

65. Loewy. Inside view of Sears Coldspot refrigerator, 1935. Raymond Loewy.

tangular plate above the door. A storage drawer for utensils was added in the base. Finally, cosmetic hardware changes completed a "redesign" that involved no major retooling costs.

The 1937 model (figure 67), according to a publicity release, was "a new landmark in modern refrigeration" owing to "its elimination of unnecessary detail" from the previous model. By getting rid of the 1936 model's chrome strip, Loewy created a "streamlined effect." Extension of sides to the floor served "to accentuate the streamline appearance" of the appliance.[14] Sears had approved a door running the full width of the refrigerator. As a result hinges were recessed and the latch-release bar yielded to a conventional handle on the door—an improvement that made it possible to close the door without grappling for a handhold or covering the front with sticky fingerprints. Finally, Loewy's office achieved "an automotive effect" by treating the entire front like a V-shaped radiator with a peak running down the center.[15] This evocation of the automobile suggests Loewy's indebtedness to GM's Alfred P. Sloan, Jr., the pioneer of artificial obsolescence through annual style changes. Despite Loewy's talk about "clean-cut and simple design" through elimination of "tricky decorative schemes,"[16] the 1938 Coldspot (figure 68) differed from the preceding model mainly in a return to an open front gap between low legs—each of which had two bands of chrome-plated tubing wrapped around it.

Since Coldspot sales increased more than did competitors', other firms and their designers extracted a significant moral from its story. But making product redesign routine did erode the romantic image industrial designers had created and maintained through devices like "signed" products. Although they profited financially from repeat performances, they no longer could bill themselves as innovators. The transition from the pre-Loewy Coldspot to the 1935 model was a true quantum leap in appearance. But succeeding models designed by

66. Loewy. Sears Coldspot, 1936. Raymond Loewy.

Loewy demonstrated only superficial modifications. How could he claim credit for introducing a radically new design when he was responsible for the one it replaced, itself introduced only a year before? Loewy might set his public relations men to writing about the evolution of the refrigerator, but that rationale collapsed after several routine model changes. Without dramatic juxtapositions of before-and-after designs, the earlier obviously created by premodernist hacks, product design gradually lost its wonder-working mystique. Manufacturers, however, had no reason to bemoan tarnishing of the industrial designer's image if he continued to provide products with the visual trappings of progress desired by consumers.

Eventually, manufacturers began to ask designers to apply to other fields techniques

67. Loewy. Sears Coldspot, 1937. Raymond Loewy.

68. Loewy. Sears Coldspot, 1938. Raymond Loewy.

that worked well with housewives. Possibly assuming that farm women helped select equipment, makers of farm implements hired designers to modernize products. In 1937 Dreyfuss began a continuous association with John Deere, while Loewy went to work for International Harvester (then McCormick-Deering) at about the same time. Unlike the domestic-appliance industry—nearly saturated with new designs —farm equipment provided limitless potential for redesign because most companies marketed products whose designs antedated the art-in-industry movement by ten to fifteen years. Design considerations in this field differed little from those in other areas. When Loewy redesigned a cream separator in 1937, his reasoning followed that of the Gestetner duplicator job (figures 69 and 70).

He replaced dangerous projecting legs with a smooth, rounded one-piece base; enclosed an unsanitary open drive belt; and removed the motor from a clumsy projecting shelf to enclosure with belt and stand in a continuous housing. For psychological reasons hygienic-looking white enamel replaced the black lacquer of the 1926 model.

Psychological considerations also entered into redesign of equipment not connected to the domestic side of farm operations. In tackling the Farmall tractor in 1936, Loewy enclosed fuel tank, engine, and radiator in a single streamlined housing, strengthened wheels, and hired an orthopedic surgeon to shape the seat, but he focused on his maxim that "a tractor, whether light or heavy in actuality, should convey an impression of solidity and strength." His design, according

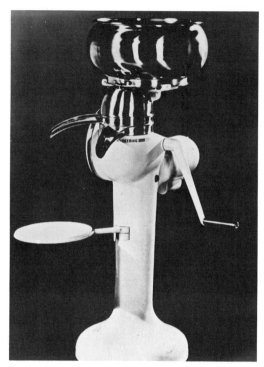

69. International Harvester cream separator, 1926. Raymond Loewy.

70. International Harvester cream separator, after redesign by Loewy, 1937. Raymond Loewy.

to a publicist, "respects the nature of the machine and dramatizes its outstanding virtues—ruggedness, dependability, efficiency —in visual form."[17] Farmers perhaps maintained more substantive criteria when purchasing equipment, but designers—and manufacturers—thought otherwise. The low point of consumer engineering for the farmer came in 1939, when the Oliver Plow Company introduced "a streamlined manure spreader," presumably the ultimate in efficient elimination of waste.[18]

It would be hard to measure the impact of style-conscious farm women on sales of agricultural implements, but women played no part in purchase of machines for industry. Yet manufacturers of machine tools and other industrial equipment also employed industrial designers. As early as 1930 the president of an arc-welding firm declared appearance "a vital factor in machine design." With the stock market crash behind it,

the machine-tool industry, like the consumer-goods industry, was suffering serious competition. Capital investments declined sharply. Customers were few. It was crucial "that machines should not only be good mechanically, but that they should have the appearance of being well built." It seemed "natural to associate the idea of efficiency with that of neat and durable appearance." Therefore, he noted, "designers and engineers in the machinery industry" were "bending their efforts toward attaining the extreme simplicity and grace of modern art in their designs."[19] Consultant designers as well benefited from this effort. Teague designed the housing of a 600-horsepower gas engine and compressor for the S. R. Dresser Manufacturing Company in 1934 (figures 71 and 72) and machine tools for the Heald Machine Tool Company in 1938 (figures 73 and 74). Dreyfuss designed a stationary diesel engine for the Fairbanks, Morse Company in 1938

71. *600-horsepower gas engine and compressor, S. R. Dresser Manufacturing Co., before redesign.* Pencil Points.

72. *600-horsepower gas engine and compressor, S. R. Dresser Manufacturing Co., after redesign by Teague, 1934.* Pencil Points.

73. *Internal grinder, Heald Machine Tool Co., before redesign.* Art and Industry.

74. *Internal grinder, Heald Machine Tool Co., after redesign by Teague, 1938.* Art and Industry.

and completed a line of turret lathes for the Warner and Swasey Company in 1939.

Some firms relied on local designers. Early in 1935 Harold Van Doren of Toledo began working on coordinated housings for the machine tools of the Ex-Cell-O Aircraft and Tool Corporation. As a young company, according to a vice-president, Ex-Cell-O sought to distinguish itself from established competitors by the appearance of its products, despite the apparent absurdity of applying industrial design to industrial goods. This official perceived four major contributions of Van Doren's work to the desirability of the firm's line. First, the new machines boosted "the pride of the operator" and thus led to better maintenance, more careful workmanship, and increased motivation for cleanliness. Second, Van Doren's careful study of problems of machine-tool use yielded modifications that increased safety and ease of operation. Third, a new line produced higher morale among Ex-Cell-O's salesmen. Devising attractive catalogues and advertisements became easier when products were attractive.

Finally, Van Doren provided a common look so that machines could be distinguished as a group from competitors. Three parallel grooves banded the base of each machine, while more compact grooves banded housings of working parts above the base.

Intersections of planes, both horizontal and vertical, were beveled. Ex-Cell-O's vice-president concluded "that the matched line of products with a common style design will become increasingly popular." With Ex-Cell-O's "radical venture" the idea of matching ensembles had come from department store to factory.[20]

Machines to Live In

By mid-decade, as Dreyfuss noted, designers began branching out from products to "the merchandising of such things as railroad tickets, telegrams, gasoline, and business, institutional prestige." They began to focus on "such intangibles as service and prestige."[21] With projects like store fronts, commercial interiors, and exposition buildings, "function" no longer signified mechanical performance but sales performance of an environment devised to increase the flow of goods or services. Form and function coalesced in a conception of industrial design as the creation of large-scale packages.

Industrial design in public transportation involved both product design and environmental packaging. Factors such as efficient use of space, lightweight construction, and passenger comfort entailed functionalism of a purist nature. But comfort, a psychological quality, also depended on creation of an ambience of modern luxury that would put travelers at ease and induce them to become repeaters. When Loewy in 1938 created interiors for Boeing 307s purchased by TWA, he hoped to provide "the feeling of an exclusive club or lounge" by means of "simplicity and crisp, modern effects." His "principal desire was to appeal to women, while giving masculine travelers textures and effects they also would like."[22]

Passenger comfort ranked first when Geddes's office prepared a mock-up interior (figure 75) for Pan American's China Clippers, which began flying to the South Pacific in 1935. Worthen Paxton oversaw the proj-

ect, setting specifications and hiring contractors, while Frances Waite Geddes did the designing. To increase comfort, Pan American authorized Geddes to reduce the plane's passenger capacity by a third. To counteract the tunnel effect of a cabin forty feet by eleven feet, Waite made four compartments connected on one side by an aisle. Three smaller compartments each contained two sofas and two chairs, while a lounge provided space for the rest of the passengers. At night each compartment slept six passengers, and the lounge converted into two dressing rooms. Geddes claimed a host of aviation firsts: sleeping facilities, a galley with range and refrigerator, separate rest rooms for men and women, hot and cold water, adjustable seats, cone-shaped ventilators and lights, sound insulation, and a shock-absorbing wine closet. Zippers made the plane's sky-blue seat covers and cream-colored wall coverings removable for washing. These accommodations did not match those of crack passenger trains, but they lent comfort to a transportation medium known then for spartan conditions.[23]

Elimination of weight entered into Teague's design of fifty passenger coaches for the New Haven Railroad in 1934. But as the company's president noted in announcing Teague's contract, "the importance of comfort and pleasing appointments" took precedence. Faced with declining passenger revenues even before the Depression, companies like the New Haven turned to industrial design to provide "a lure to win back the business lost to airplanes, buses and other forms of transportation." In search of a design that would "embody a close approximation of current consumer demand," the company proudly announced that it had "broken a tradition of the industry" by choosing "a leading designer of consumer products to satisfy the public's desires in the new coaches."[24]

Streamlining would soon become a national preoccupation. Anticipating this, the New Haven hoped to attract travelers with

*75. Geddes. Women's lounge in Pan American
China Clipper, 1935. Geddes Collection.*

modern streamlined coaches despite
Teague's belief that aerodynamics had few
functional contributions to make to railroading. Trains were too long for bulbous locomotives and tapered observation cars to reduce wind resistance. According to Teague,
a tubular cross section might reduce cross-
wind friction. Even so, the principal gain
would be increased fuel efficiency rather
than the increased speed expected by the
public. But he strove to produce coaches
that would *look* streamlined. Backed "to the
limit" by the New Haven, he fought the
"difficult, hidebound, obstructive" engineers
of the Pullman Bradley Company of Worcester, Massachusetts, which manufactured the
new coaches. Against objections he spec-

ified a car thirteen inches lower than nor-
mal with a semitubular body reinforced
on the inside rather than the outside, no
exterior moldings, hinged skirts concealing
wheel trucks, and welded exterior construction to replace traditional but visually distracting rivets. He compromised on retention of rivets because he "became convinced
that it really would be impossible for Pullman Bradley, at least, to use other methods."
Windows were grouped in pairs and outlined with polished aluminum "to emphasize
the horizontal lines of speed." Painted bright
green, the new coaches presented an exteri-
or "much cleaner and neater in appearance"
than the high, square, molding-encrusted
body of the standard passenger coach.[25]

*76. Teague. Rendering of interior of New Haven
car, 1934.* Pencil Points.

Teague devoted even more care to the
coach interior (figure 76). A flashy streamlined shell might attract the curious, but to
retain passengers the railroad had to deliver
the goods. He coordinated his interior design to produce an "effect . . . of comfort,
cleanliness, coolness and gaiety." Above all,
he concentrated on replacing the overstuffed
ponderousness of typical coach design with
a light quality expressing his decade's emphasis on efficiency, its belief that less is
more—nearly a philosophical concern that
was marked in design by use of unadorned
materials, particularly light metals, alloys,
and synthetics, in their pristine machined
state. Borrowing from European design,
Teague provided tubular aluminum seats and
removable cushions. He determined "the
proper slant of back and height" of the seat
by having "fat, thin, short, tall" individuals
test several prototypes. A circular steel shaft
supporting each seat left legroom and space
for belongings. Molded Bakelite armrests
did not provide much comfort, but they
added to the machined beauty of polished
aluminum tubing. Luggage racks constituted
"an unhappy necessity from the point of
view of design," but he compromised with a
slender grill of extruded aluminum.[26]

Interior colors reflected a metallic iciness.
Dark gray-blue floor, upholstery, and lower
walls yielded to silver walls above the lower
window sills and an off-white ceiling. Teague
found it necessary to "counteract any effect

of coldness" by running three parallel bands of "brilliant vermillion" near the top of each wall—a flash of speed. He justified use of bright colors by arguing that air-conditioned coaches did not need dull grays and browns traditionally used to nullify effects of ashes, soot, and grime. This naive faith in local environmental control reflected industrial design's larger goal—elimination of frictional irritants and uncertainties from a totally controlled environment. If the New Haven's new coaches made traveling less irritating by providing pleasant, clean, spacious machine-age accommodations, they also relieved the economic uncertainty of the railroad's passenger operations. Attributing increased revenues to Teague's work, New Haven officials ordered fifty more identical cars a year after delivery of the initial lot.[27]

Raymond Loewy's office took on a more complex job in 1936 when the Panama Railway Steamship Company commissioned him to design interiors of three sister ships, the *Panama*, the *Ancon*, and the *Cristobal*, eventually launched in 1938 and 1939. According to a press release, recent ships of other nations, particularly the *France*—outfitted in the exposition style—exhibited "modern interiors representing the particular decorative trend in their respective countries." But American shipbuilders still rigorously "followed classical styles, such as Colonial, Queen Ann [*sic*] or Regency, or adaptations of such styles." Loewy did not object to historical styles in the privacy of one's home; he even admitted, "Two of my own are decorated in period style." But public transportation interiors should reflect the "SIMPLICITY" of a new national "type of decoration" found throughout the United States. They should reflect "Contemporary American" style. Therefore, he provided the three ships "with smart, appealing and up-to-date backgrounds—interiors that suggest the casual and informal atmosphere of the modern club, hotel, or lounge."[28]

Interiors of these ships reflected propensities common among "machine-age" design-

ers and architects. New fire regulations dictated use of "new metals and alloys, plastics and synthetics, finishes and textured surfaces"—the "unlimited mediums of expression . . . today at the disposal of the creative artist," as Loewy's publicist characterized them.[29] Stainless steel doors, metal tables with Formica tops set in aluminum molding, chairs with squared tubular frames supporting spare upholstery, polished metal banisters with rounded corners and geometric lines, and exposed structural beams in ceilings—all proclaimed a steamship as a machine for living in temporarily. Colors of walls, upholstery, and carpeting softened the metallic tone of his interiors but contributed all the same to an effect of impersonal neutrality. Loewy hoped that muted tones would calm passengers' seagoing anxieties. Along this line he specified "flesh-tinted mirrors" that would counteract "a tendency toward mal-de-mer" by enabling woozy voyagers to "view themselves as healthy, sun-tanned individuals."[30]

Desire to make passengers comfortable—to make them feel at home—inspired some departures from the common machine aesthetic of the thirties. Otherwise spare, functional cabins had window curtains emblazoned with bright multicolored horizontal stripes. Doors in the main hall (figure 77) bore meretricious ornament—vertically aligned boxes within boxes outlined in relief with aluminum. A rectangular clock face resembled a plastic clock radio of the fifties. Superficially streamlined by elimination of numerals, the clock had such a horizontally distorted shape that it appeared delicate, even precious. These and other self-contained decorative touches were incompatible with the general machine-age tone of his interiors and contributed a jarring busyness, especially when compared with an integrated interior like Geddes's Elbow Room restaurant of 1938, later renamed the Barberry Room (figure 78). The point here is not Loewy's personal taste but that he made ship interiors homelike by including details

77. Loewy. Main hall of the Panama Railway Steamship Co.'s liner, Panama, *1938. Raymond Loewy.*

78. *Geddes. Elbow Room (renamed Barberry Room), New York, 1938. Geddes Collection.*

he would have liked in one of his own modern homes. Domestic taste rarely concerned Loewy when he, like other designers, worked on landlocked business and commercial interiors and exteriors. For public facilities he retained an impersonal tone thought conducive to the efficient conduct of business, the selling of goods and services, or the projection of favorable images.

In 1935 Loewy told a reporter that industrial design would expand from consumer products to airports, transportation terminals, and business centers.[31] In *Horizons* Geddes had provided ambitious plans for two airports: one a floating, rotating structure to be located in New York Harbor below Battery Park, the other a design for any medium-sized city that would have saved time by eliminating private aviation, separating passenger and freight operations, and servicing a plane a minute with an automated system of luggage conveyors, escalators, and moving sidewalks.[32] But public projects exceeded the scope of industrial designers, who were limited to private

patronage. Designers' architectural commissions originated in the offices of corporate executives who manufactured or sold their redesigned products. Van Doren, for one, thought that modernization of commercial fixtures like drinking fountains, gasoline pumps, vending machines, office equipment, and grocery-store coffee mills inexorably followed redesign of consumer products and packages.[33] The same could be said of modernization of commercial interiors, whether offices or retail outlets. To avoid disharmonies, the corporate display room where a new product was "sold" to distributors and the store where it was sold to consumers had to express the modernity embodied in the product. Or, at least, stylistic harmony between product and surroundings made the sales process easier. Thus, products and sales environments perpetually evolved in relation to each other as they sought a stylistic equilibrium that remained elusive.

Designers provided redesigned equipment as well as plans for offices, display rooms, and stores. Office modernizers could choose from metal desks designed by Dreyfuss for Yawman and Erbe—at the top of the line an option-loaded $1,000 model intended to dispose company presidents to equip entire offices with less expensive models.[34] Executives in search of new business machines could choose from those redesigned by Teague in 1936 for the A. B. Dick Company, and they could shop in a Chicago showroom, also designed by Teague, with an exterior (figure 79) of black Carrara glass trimmed with aluminum and a sleek interior (figure 80) of wraparound Flexwood paneling set in satin-buffed, chrome-plated moldings.[35] Retailers as well could upgrade surroundings with Teague's WIZ autographic sales receipt register (figure 57) or his new machines for the National Cash Register Company.

A Californian protested that National Cash Register's new trademark—"NCR" in block letters replacing the firm's full name in ornate script—seemed devised by "second or third class brain trusters" embarked on formation of a "New Communistic Republic," but most customers, whether businessmen or consumers, embraced modernization as an optimistic expression of confidence in a difficult time.[36] At least the volume of redesigned offices, showrooms, and stores indicated that they did. All the major industrial designers but Geddes contributed corporate offices and showrooms in the thirties, and he had led the way with the J. Walter Thompson auditorium.

Dreyfuss designed showrooms for the Dennison Manufacturing Company (1930), the National Blank Book Company (1930), Chrysler (1936), and B.V.D. (1937). Loewy provided the United States Plywood Corporation with an office (1939, figure 81) designed to show off its products as finishing materials. And Teague, most prolific in this area, designed display rooms and offices for Eastman Kodak (1930, 1931, 1934, 1936), A. B. Dick (1935, 1936), the Bryant Heater Company (1936), the Crosley Corporation (1937), the Square D Company (1937),

National Cash Register (1938), the Universal Atlas Cement Company (1939), and the Ford Motor Company, which used his services from 1933 to at least the close of the decade. These examples, drawn from the leaders of the profession, represented a fraction of the total number of projects carried out. Countless firms hired less well-known designers, architects, or interior decorators to provide surroundings in harmony with the times and with their own modernized products. An office with dark wainscoting, carved railings, gilt lettering, cuspidors, and white globe lights now signified a company mired in Depression doldrums.

Providing an environment to impress distributors and retailers supported only the first step in selling a product. More significant was the environment in which a consumer made purchasing decisions. A survey conducted in 1939 by *The Architectural Forum* revealed that from 1924 to 1938 three-quarters of the nation's commercial establishments conducted face-lifting operations. Of these, 49 percent modernized once, 29 percent twice, and 21 percent a remarkable three times.[37] Industrial designers participated in only a tiny fraction of these jobs. But as visible leaders they influenced the forms of new facades and interiors.

Through an association with the Pittsburgh Plate Glass Company, Teague contributed more than any other designer to a common style for commercial facades. Retained in 1935 and 1936 for research and development, he worked on applications of glass to furniture, interiors (primarily glass bricks), and store fronts. After going through the firm's plants for manufacturing plate glass, structural Carrara sheet glass, and metal fittings, Teague concluded that "these materials . . . are plastic and versatile beyond anything I had imagined" and set about publicizing their use.[38] By October 1935 his draftsmen had prepared twenty-eight renderings of various store fronts with glass as primary facing material. Four months later he reported completion of twelve more de-

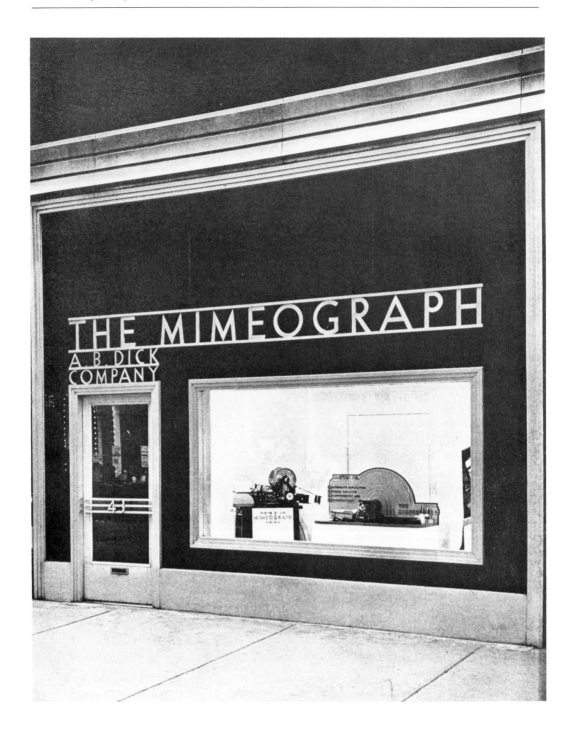

79. Teague. Showroom for A. B. Dick Co.,
Chicago, 1936. Architectural Record.

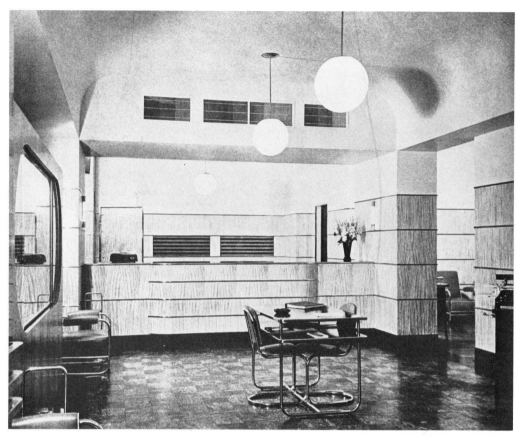

80. Teague. Interior of the A. B. Dick showroom.
Architectural Record.

signs on which his staff had "given special
attention to new and unusual lighting effects
as well as to new and unusual applications
of carrara glass."[39] These prototypical store-
front designs included all kinds of commercial establishments, from taverns (figure 82)
to high-fashion clothing shops.

Essentially, Teague was engaged in two-
dimensional poster design—creating visual
patterns from basic elements of doorway,
display window, sign, baseboard, wall, and
molding. Typically, these facades included
Carrara baseboards of black or deep violet;
contrasting walls of light-colored opaque
Carrara (often green, yellow, or cream);
boldly cut stylized letters of dark molded
glass projecting from the surface; and horizontal accenting stripes of chrome or con-

trasting glass. Recessed entrances invited
customers with curved display windows and
baseboards, flowing inward on each side.
Most of Teague's facades projected a polished appearance of smooth precision and
geometric regularity, an effect heightened by
visible joint lines of rectangular Carrara
sheets. These store fronts did indeed exem-
plify "design for the machine," as the lettering proclaimed on a sample commercial
facade that Teague constructed for a 1932
industrial design exhibit at the Pennsylvania
Museum of Art.[40]

No building actually received a Teague
facelift. Pittsburgh Plate Glass commissioned
his designs for publicity. The renderings
appeared in advertisements and promotional
brochures. In 1936 the company mounted a

81. Loewy. Bar area of office of United States Plywood Corporation, 1939. Raymond Loewy.

traveling exhibit in a truck designed by Teague and equipped with cleverly lit scale models of several of his store fronts and samples of architectural glass.[41] Well-packaged, his ideas reached merchants across the country who were seeking new appearances for their own goods and services. Teague's work probably had significant impact on the appearance of main-street America. Since more architects and contractors ordered structural glass from Pittsburgh Plate Glass than from any other supplier, it seems reasonable to assume that they followed the company's suggestions for using it. In any event, facades similar to Teague's sprang up everywhere during the late thirties, often oddly superimposed on distinctly premodern buildings. Such stores appear today in urban shopping districts that have been unfashionable for years, like St. Louis's Gravois Boulevard, or in countless small cities, like Marquette, Michigan, or Derby, Connecticut, where modernization is a thing of the past. Now part of the general urban blight, with their Carrara cracked and their glass bricks covered with grime, they once signified economic recovery to merchants who invested in them and attracted customers with their machine-age gleam.

While Teague was thus acting as publicist, Loewy's office actually became involved in renovating old stores and planning new ones. In 1937 his office remodeled a corner bakery operated in New York City by Cushmans Sons (figure 83). Intended as a prototype for the company's chain of three hundred stores, the design had to project an image easy to recognize. This store could have served as one of Teague's Pittsburgh Plate Glass renderings, but Loewy used white sheets of porcelain-covered steel instead of glass as a finishing material. Dark cornices and baseboard provided contrasting horizontal accent. Streamlined display windows terminated in a half circle at each end of the facade. As in some of Teague's designs, one wall curved inward to meet the door, which had a typically round window. The exterior, too high and square for the speed Loewy sought to project, also suffered from ornate script lettering and a pennant-like sign suspended from a projecting pole—trademarks the company refused to discard.

Within the store (figure 84), Loewy got rid of rectangular wooden display cases and baseboards with decorative molding. He provided instead a single display counter sweeping the store's length and ending in a full curve. Entirely of glass, it had fittings not of the usual chrome but of brass, whose "warmer tones," according to one account, "blend more harmoniously with the colors of bakery products."[42] But incongruous details nullified this streamlined effect. A functional white porcelain wall on the proprietor's side of the counter maintained an

83. Loewy. Exterior, Cushmans Sons bakery, New York, 1937. Raymond Loewy.

uneasy relationship with gold-figured reddish-tan wallpaper on the customer's side. Decorated with a large round mirror, a framed picture, and a small urn on a camouflaged heater, this wall projected a disconcerting domesticity. A dark blue linoleum floor contrasted with the hygienic serving area. This expanse of blue was interrupted by a white arc from door to counter and a white circle marking where a customer would stand to pay for purchases. Despite this interior's unassimilated elements, Loewy used motifs then emerging as hallmarks of streamlined architecture—curved fixtures and flow lines intended in commercial structures to lead consumers without obstruction to the point of purchase. Such motifs ex-

pressed visually their intended commercial function.

Two interiors designed by Dreyfuss in 1936 embodied similar intentions with greater subtlety. He provided Western Union with a telegraph office in Philadelphia (figure 85) that was to be a prototype for modernization elsewhere. *Modern Plastics* claimed that it pointed "the way to a fuller usefulness of a store from the standpoint of advertising and sales promotion." Dreyfuss himself noted that "because speed is the essence of the business of the telegraph company," he and his staff tried "to capture a feeling of speed and efficiency and to express them in a pleasing modern design."[43] Curved lines predominated. The sides of a

84. Loewy. Interior, Cushmans Sons bakery, New York, 1937. Raymond Loewy.

long, low indented service window curved into the customer lobby's semicircular rear wall. Above this window a clock's yellow plastic numerals and aluminum hands projected directly from the deep blue wall— whose color contrasted sharply with the businesslike yellow of the employee area beyond. In the center of the lobby stood a circular writing table, surmounted by a slender circular blue column with a frieze of Mercury figures in gold. A ceiling of yellow —the color of a telegram form—and a floor of linoleum patterned with one-directional slash lines of various colors—intended to recall staccato bursts of telegraphy—completed the design. The office's predominant lines evoked a mood of smooth, fast, un-

broken continuity. And its pleasant color scheme, Dreyfuss hoped, would overcome the public's traditional association of the telegraph with bad news and would condition its use for informal communication.

A second Dreyfuss interior expressed industrial design's central function in architecture during the thirties—promoting the flow of people and images. Designed with architect Julian G. Everett, a member of his staff, the Socony-Vacuum Touring Service Bureau (figure 86) occupied a street-level space in Rockefeller Center's RCA building. The oil company (now Mobil) called in Dreyfuss because the previous bureau, whose main purpose was not dispensing maps but publicizing the company, had not attracted

85. Dreyfuss. Western Union telegraph office,
Philadelphia, 1936. Dreyfuss Archive.

enough people. The new bureau occupied a
U-shaped space with street entrances on the
curved end and an interior entrance on the
straight end. Two curving windows enticing-
ly previewed exhibits within. Between windows the interior wall curved in to form a
U-shaped projection concealing a structural
member that had marred the previous arrangement. Thus, the interior wall formed
two sinuous S-curves, one a reflection of the
other. Around the bureau's flowing walls
were various displays—illustrations of petroleum products, large electrified relief maps
showing locations of Socony operations,
photomurals, continuously running movies,
and push-button models of gasoline engines.
Wall contours led visitors easily from one
display to the next. Polished railings, a band
of indirect lighting at intersections of wall

and ceiling, and projecting captions in metal
letters ("AROUND THE WORLD WITH SOCONY
VACUUM") added to this flowing continuity.[44]

More was at work here than sophisticated
traffic control. The smooth flow of images in
displays was not interrupted by complex
information that would provoke thought.
Visitors marveled at the gadgetry used to
present what little information there was.
Entranced by technological wizardry unrelated to the company's fields of expertise,
they took from their visit to the bureau an
awareness of Socony's dedicated contribution to modern life. These techniques of
manipulation—traffic flow, gadgetry, and
presentation of superficial images—typified
application of consumer engineering to interior spaces designed for selling prestige
rather than products. Stripped of material

content, no longer even packages in the strict sense, designs like Dreyfuss's touring bureau functioned purely as commercial lubricants, easing sales resistance.

Lubrication was also an aspect of designers' more permanent architectural endeavors. Teague and Geddes provided oil companies with service station designs. Their involvement made sense for several reasons. First, since one brand of gasoline differed little from the next, product differentiation by extrinsic means became essential. Second, to establish brand loyalty, all of a company's stations had to present an easily recognized, uniform appearance, derived from a prototype as carefully planned as a clay refrigerator model. In theory, designing a gas station was design for mass production. Finally, oil companies profited considerably by selling lubricants, parts, and accessories. These had to be packaged to reflect a uniform corporate image and arranged in eye-catching displays. Product design work provided the experience for satisfying these requirements. Similar considerations later transformed the American environment with standardized supermarkets, convenience stores, motel chains, and fast-food franchises. Lubrication of the path to the door of such establishments physically—through easy access and fast service—and psychologically—through pleasant surroundings and repetition of visual motifs—became the primary focus of service-station design.

Teague's fruitful association with Texaco led, between 1934 and 1937, to several variations on a basic station design. Robert Harper of Teague's staff supervised the project, which began with a survey of operators and customers at about twenty service stations. Most drivers picked a station because of its cleanliness, others by brand recognition, and only a few took the first available. Only a fraction entered the station building during servicing. Station operators stressed the importance of seeing all pump islands and driveways from inside the office. To meet these conditions Harper and chief

draftsman Hopping designed a station with large glass areas, white walls of easily cleaned, porcelain-covered steel sheets, and displays of accessories at each pump island. Provisions for customer comfort included canopies over pumps, removing services like battery charging from waiting rooms, and adequate rest rooms. For a while Texaco inspected its "Registered Rest Rooms" with a fleet of forty-eight cars designated the "White Patrol." This ploy drew some customers, but the stations' distinctive appearance attracted more.

The most common type was designed to sit back from a highway on a large lot. This "Type C" station (figure 87) was rectangular, had white walls, two service bays marked "LUBRICATION" and "WASHING" in projecting red block letters, and an office whose expanse of glass was framed in green and set off by a green baseboard. A flat streamlined canopy—its edge accented with three thin green stripes that continued around the building—extended from the office to cover the pumps. On each side of the canopy stood a large red "TEXACO" sign in three-dimensional block letters against a streamlined white backdrop. Two small plain red stars were located above the striping on the service area and on each side of the station. By 1940 Texaco had built five hundred such stations and remodeled others to meet the general design. Neat, clean, dignified, but colorful in a restrained way, the station proved so successful that many remain, unchanged, particularly west of the Mississippi (see figure 88). Although the canopy was mildly streamlined, Teague knew when to avoid passing styles for a neutral and thus timeless design.[45]

When Geddes designed a prototype for Socony-Vacuum in 1934, he also insisted that a service station should be "simple, attractive and neutral" rather than modernistic. Instead of relying "for its effect on fantastic forms and shapes," its appearance should be "primarily based on utility and economy" to "create the impression of modern, efficient

86. Dreyfuss. Socony-Vacuum Touring Service Bureau, New York, 1936. Dreyfuss Archive.

*87. Teague. "Type C" Texaco service station,
1934–1937.* Architectural Record.

*88. "Type C" Texaco station as it stands today,
Sulphur Springs, Texas. Author.*

service."[46] Socony built only one experimental station based on Geddes's design, but a long report submitted to Socony revealed factors Geddes found significant in service-station design. This report followed "an analytical study of service station functions and characteristics" with specifications for an ideal station. Geddes hoped to increase gasoline sales through "greater attractiveness, distinctiveness, customer comfort and merchandising force," even though, as he observed, he intended no improvements in the technique of dispensing gas. Meticulous planning of "psychological factors" constituted his main approach to the problem. Literally and figuratively, he intended his design to provide friction-reducing lubrication. In fact, his report, filled with close rationales for the slightest details, is a remarkable primer on applied consumer engineering.

First, according to Geddes's report, one had to attract motorists' attention. Earlier Socony stations vaguely followed colonial architecture, sometimes brick, sometimes wood. But Geddes insisted that all stations should follow a prototype "so distinctive that it will be in effect a Socony trademark." Ideally, no sign would be necessary. But he recognized such a departure as impractical and settled in his design (figure 89) for a slender pylon above each pump island, bearing the word "SOCONY" in vertical letters on each of three faces. A station building should sit diagonally at the back of a corner lot, a little off center to avoid symmetry. This arrangement would provide space for islands to be located far enough apart from the building and from each other to permit a car to reach any pump without backing up or turning around. "Ease of entrance," Geddes maintained, "is vastly more important in attracting customers and a certain amount of exit ease can profitably be sacrificed." Flow lines painted on the pavement would indicate routes from entrances to all pumps and bays. With lubricated traffic flow assured by distinctive appearance, accessible

pumps, and flow lines, Geddes moved on to consider lubrication of the motorist himself and his automobile.

"At every point" in the station design, he noted, "an attempt has been made to impress the need for lubrication on the customer, to offer simple access to the space, to give the customer greater comfort during lubrication, and to develop the character of the lubrication process so that it will be more attractive, interesting, educational and dramatic." He hoped "to make the service a stimulating experience, in contrast to the usual practice of treating it merely as a casual necessity of motoring." Almost an obsession, this emphasis on lubrication dictated many facets of his station design. Many stations serviced cars in open air or under a canopy, but Geddes suggested enclosing service bays with the office both to indicate lubrication's importance and to make the building larger and more impressive. His design included two bays taller than the rest of the building, equipped with glass doors to expose work to casual motorists and accented by curved walls flowing in to meet them, one of which necessitated a small three-dimensional "false front." He advocated using lifts instead of less expensive pits so that the "lubrication procedure may be a graphic, interesting, and educational spectacle" for customers, who could view it through a glass partition separating the waiting room from the first service bay.

The pleasant atmosphere of the waiting room also promoted lubrication (see figure 90). Circulation patterns kept station employees from barging through the room, which was separate from the office because "the customer has little comfort if traffic constantly passes through the room, and he can have no sense of privacy." Employees reached the men's room through a rear door hidden from female patrons, whose separate facilities were protected by a baffled entrance and then by a powder room. Treated with respect and securely settled in the cushions of a tubular chair, a customer could

*89. Geddes. Service station model for
Socony-Vacuum, 1934. Geddes Collection.*

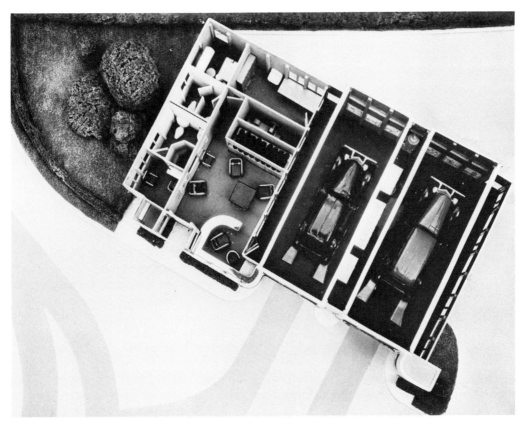

90. Interior of Socony-Vacuum model. Geddes Collection.

watch his or her car being serviced or yield to "the subtle suggestion of attractive . . . merchandise and educational material" displayed on a streamlined partition between waiting room and office.

By easing the way into the yard and thence into the station itself, through attractive appearance and easy exterior and interior traffic-flow patterns, Geddes hoped to get a customer's car up on a lift for lubrication. With the car suspended in the air, its vulnerable belly exposed, he or (especially) she would be "in the most receptive frame of mind" and "most easily influenced by suggestion" to agree that further work was necessary to put the car back on the road. Every element of the design addressed elimination of customers' natural resistance to having cars put on a lift. His "efficient, orderly,

attractive and inviting" solution would provide total control of a gas station's environment. In his zeal to eliminate every uncertainty in advance, he specified in his design "definite facilities for all services, materials, and equipment, leaving nothing to chance disposition or personal choice of the station manager and attendants." They, like customers, would become necessary but replaceable parts of a process organized by consumer engineer Geddes.

Industrial design's scope never extended beyond projects such as these gas stations. Their isolation as mass-produced units reflected the private enterprise system that made them possible. Teague suggested in 1935 that product-design methods be applied to community planning, and the enthusiastic Cheneys claimed that "everything from a bread box to a city *is* architecture—and industrial design." But such statements did not gain control of, or even entry into, a design field so broad that it remained the province of politicians, power brokers, and usually powerless public commissions.[47] Until World War II brought them defense contracts, industrial designers remained dependent on business patronage. Their influence extended no further than the interfaces of buyers and sellers or consumers and producers. When designers did become involved in community planning, their approach remained private.

In 1937 Teague's office provided United States Steel with studies for a housing development to be located near Pittsburgh.[48] And in 1934 Geddes presented Nelson A. Rockefeller with a plan for rescuing Forest Hill, an upper-income development outside Cleveland that had failed early in the Depression. Anticipating later subdivisions, Geddes recommended lower-priced apartments and houses, wide choice of exteriors, winding streets with frequent dead ends, footpaths connecting dwellings to schools, and built-in kitchen appliances. Geddes told Rockefeller that he considered the project "an important social contribution as well as in terms of brick and mortar." Rockefeller himself, declaring the report "a brilliant piece of work," could not recall "when I have seen a more comprehensive or beautifully prepared study." Despite Geddes's honest dedication to "social planning," however, the crucial factor in planning the project remained development of "stimulating psychological factors that the younger generation are looking for in the houses they buy."[49]

These projects did not reflect industrial-design method nearly as much as did the movement for mass production of prefabricated housing, in which Teague and Geddes became involved.[50] Making a house in a factory transformed it from architecture to industrial design and opened a way, as the subsequent history of mobile homes re-

vealed, for superficial styling changes. But during the Depression prefabrication seemed ideal for housing thousands of uprooted people. In 1936 Teague predicted that in ten years "a modern functional type of domestic design" would replace "costume pieces . . . very poorly adapted to modern living conditions, modern domestic equipment and modern methods of construction." The public would recognize the "superior beauty" of the "'machine to live in,'" just as it had accepted superiority of "a modern streamlined automobile" to an "imitation of horsedrawn carriages."[51] When Teague returned from meeting Le Corbusier in Europe, he brought home the master's suggestion that he convince Edsel Ford to apply the assembly-line principle to domestic architecture.[52] Ford expressed little interest, but by 1940 Teague was seeking corporate and government funding for a demountable house of plywood panels to be equipped with a one-piece factory-made utilities unit. Only "through the industrial production of a house" could the nation complete the "environmental reconstruction" that constituted its "most urgent problem."[53]

But Teague's project came to nothing. The outbreak of war intensified a need for housing but absorbed a motivating fear of social unrest implicit in Teague's appeal. Geddes's similar project, begun in 1939, ended two years later when his sponsor, the Housing Corporation of America, went bankrupt. The Revere Copper and Brass Company then picked up Geddes's design for a mass-produced prefabricated house for use as a promotional gimmick. Magazine advertisements featured photographs of a model of his design and lauded the project for envisioning future applications of metals.[54] Thus ended industrial design's involvement in development of architecture for the masses. Teague's plywood box, actually a far cry from the elitist "machine to live in," failed to interest business once the threat of social unrest and the promise of government subsidy had dissolved. The fate of Geddes's project also demonstrated the lack of influence of the industrial designer divorced from corporate patronage.

The extent to which a designer departed from his role of lubricating the flow of goods and services from producer to consumer indicated the extent to which he had entered a realm of unreal images—useful only for publicity. However much designers like Teague and Geddes preferred to see themselves as social engineers designing and controlling a machine-age environment, their control extended only to areas conceded by corporate patrons. Since they lacked political power, direct social control remained an illusion. But they celebrated their illusion—its potentiality split among scores of competing firms—in the frictionless surfaces of the streamlined style. And although full social control eluded them, statistical control of countless points of contact between consumers and producers did not.

The Practical Ultimate

Such criteria as "form follows function" are unsound because ... they make the error of assuming that there is one, and only one right way of doing a thing.
—J. F. Barnes and
J. O. Reinecke

Even idealistic industrial designers admitted that the profession existed to make products attractive to the public, or to that segment able to buy them. By definition designers had to appeal to the masses. A designer's aesthetic seriousness thus depended on his assessment of public taste. Conversely, the quality of taste consumers could express depended on available goods. By 1948 Siegfried Giedion found that industrial design equalled Hollywood in "the shaping of public taste." Reliance on the merchandiser as a "dictator of taste" was "a source of danger and bondage."[1]

By the end of the thirties many designers became as manipulative as Giedion later implied they were. Harold Van Doren's *Industrial Design*, a 1940 textbook, advised assessing "*standards* of taste" for a product's market as a first step in the design process. Through experience he had found that factory workers purchased plain washing machines but preferred radios and automobiles encrusted with "borax." Rather than maintaining the highest aesthetic standard, according to Van Doren, whose own work possessed aesthetic merit, a designer should choose a "middle course." Without underestimating the taste of the public, he should provide "the very best it will absorb, and not one bit more."[2] In 1946 Loewy even questioned the idea of beauty. Thumbing his nose at British purists, he told the *Times* of London that Americans had "far too many other good reasons for using" industrial design than "aesthetics"—in particular "increased sales appeal of products and increased trade." To an American designer, he claimed, a "conception of aesthetics consists of a beautiful sales curve, shooting upward."[3]

Few American designers had so blatantly dismissed aesthetics ten or fifteen years earlier. Only Guild proclaimed the need for "styling down" products to meet public taste and bragged of success in doing so. Joseph Sinel countered current opinion by declaring that "the mass responds readily to gaudy, tricky things" because it "is Coney Islandminded." Unlike Guild, he thought responsible designers were educating the public to appreciate valid design.[4] Most designers and their publicists subscribed to a designer-as-educator rationale. The designer's public responsibility included bringing to consciousness the average person's instinctive appreciation of beauty. As popularizer of styles developed in the fine arts, the industrial designer had an opportunity to raise the general level of taste in the country.

Toward the Style of the Century

Designers frequently claimed their work would improve public taste and spark a cultural renaissance. A vigorous proponent of this view, Teague predicted that as "the streets and houses, shops and factories" filled with objects of beauty, "our eyes may become so accustomed to everyday beauty in time that we may learn to appreciate painting and sculpture again, and the museums and picture galleries may cease to be such lonely places."[5] Although a realist like Loewy might caution the designer to use "simple language" because "he is addressing himself to the masses," most designers publicly accepted a role as civilizers.[6] Geddes concluded that industrial design "would slowly but surely have a constructive affect [*sic*] on the opinion and judgment of . . . masses of people."[7] With the president of the Worcester Pressed Steel Company, designers believed that "the widest field for the dissemination of culture and beauty is in the industrial products of this machine age."[8] By bringing modern machine art to the masses, they hoped to catalyze a cultural renaissance as well as economic recovery. Together, these goals imbued the industrial design movement of the early thirties with nearly messianic fervor.

Many designers considered themselves to be working at a unique historical moment. They understood a machine age as a time when the human race was becoming con-

scious of the machine's predominance in all areas of life. As Geddes noted in *Horizons*, humanity had so far been victim rather than master of the machine.[9] Other designers shared a negative assessment of the industrial revolution that reached fullest expression in Lewis Mumford's *Technics and Civilization* (1934). Mumford described the "paleotechnic phase" as a necessary "disastrous interlude" marked by "growth and multiplication of machines"—accompanied by exploitation of workers, factory towns that bred disease and alienation, air and water pollution, and, most significant for designers, ugly machine products. According to Mumford, the end was in sight. Industrial society was undergoing a "mutation," entering the "neotechnic phase." Heralded by clean electric power, light alloys, synthetics, and automation, this new age would witness "the refinement, the diminution, and the partial elimination of the machine." Society approached a "dynamic equilibrium" marked by "balance, not rapid one-sided advance: conservation, not reckless pillage." The previous phase of industrial expansion had "helped by its very disorder to intensify the search for order, and by its special forms of brutality to clarify the goals of humane living." Freed from the need to sacrifice themselves to their machines, people would have leisure to organize a harmonious society devoted to conservation, economy, and efficiency. The machine, once a disrupter of social order, now permitted rational design and control of a social order as perfect as human nature could devise.[10]

Among designers, only Geddes and Teague advocated social design by arguing that industrial design principles retained validity in broader applications. But even less expansive colleagues discussed ways to overcome the industrial revolution's unfortunate impact on design. Their common analysis revolved around the divorce of art from industry as machines replaced craftsmen. As Teague noted: "New technologies followed so fast upon each others' heels that we didn't even have time to master them, much less for that patient experimenting, characteristic of our ancestral craftsmen, in the perfecting of forms." As a result, "we could do nothing but dig in and hold fast to the only unchanging thing in the aesthetic world—the Art of the Past."[11]

When they considered beauty at all, manufacturers had relieved the naked ugliness of products by scattering fig leaves over them. The nineteenth century had witnessed steam engines framed by Corinthian columns, mechanical saws whose stands boasted intricate iron ivy, and building fronts of metal imitating stone, brick, and wood. This guilty veneering of machines and their products culminated in the World's Columbian Exposition of 1893. Its machinery exhibits, intended as a monument to the industrial revolution, filled a classical building complete with entablature, colonnades, domes, and sculptures. The exposition's ensemble of classical structures constituted a whitewashing of the brutal realities of life during industrialization.

Engrafting historical styles and outdated forms onto machines and their products hardly ended at the turn of the century. According to reform-minded industrial designers, new machines like autos and radios—with forms not established by tradition—suffered most from beautification attempts. They tended to resemble objects that earlier served similar functions. The automobile for years was truly a horseless carriage, while the radio console took its lines from the victrola, in turn derived from a cabinet used for storing sheet music. Designers like Teague hoped to overcome a legacy of "wholly irrelevant ornament which had nothing whatever to do with the product or its use."[12] By 1930 he perceived the "chaos of transition" yielding to "an unmistakable focussing of tendencies," as designers evolved "a new vision" that would finally, after more than fifty years of aesthetic clumsiness, encompass "a Machine Age, an Age of Power, an Age of Mass Produc-

tion."[13] This vision, another observer noted, required understanding of functionalism "to relate artistic design to modern materials, current social needs, and present-day industrial processes."[14] By bringing together form and function, the industrial designer would renew the marriage of art and industry.

Tutored by Europeans like Le Corbusier, American designers looked at domestic artifacts that had escaped fig-leaf beautification for examples of functionalism. Utilitarian objects—suspension bridges, grain elevators, factories, even kitchen appliances and bathroom fixtures—had escaped decorative elaboration because they were not intended for show.[15] As designers grappled with functionalism, they came to regard even the modernistic style—once expressive of the machine—as reprehensible ornamentation. According to Dreyfuss, the word "modernistic" described "odd-looking objects thrown together to look extreme and outlandish," unworthy with their "unnecessary frills and fancies" of "the lowliest speakeasy." Loewy considered "modernistic" design "little more than an angular treatment of the opulent Victorian." And Geddes, who found in the "modern*istic* style" the "applied decoration" of "fussers, gadgeteers, pseudo-artists," suggested that a "sincere style" would be "the direct result of the problem to be solved, of the materials involved, and of the sensitiveness of the artist."[16] Most designers agreed on what constituted bad design, but a definition of good functional design proved harder to come by.

A clear statement on functionalism came from Sidney G. Warner of the art division of Westinghouse's engineering department. According to him, the contemporary designer's "creed"—"form follows function"—had two corollaries: "An object should not disguise the basic principle of its construction"; and "it should not disguise the materials of which it is made." Geddes echoed this interpretation when he declared that to achieve "sincerity and vitality" in a design, its "surface qualities" should be "the direct expression" of its "structural qualities." Teague likewise found "*candor*" to be "*the simple secret of true modernism in design.*" According to him, "the things we make" have to "be candidly expressive of the materials and methods used" in construction. But they also have to be "candidly expressive . . . of the purposes for which they are intended."[17]

Teague would have argued that a definition like Warner's was sufficient to the extent that it eliminated meretricious ornament. It suffered, however, from its attention to an object as a static thing when, in fact, most designers worked with machines or products whose nature was revealed only in use or, as with the automobile, in motion. Dreyfuss addressed this problem of function in motion—somewhat different from functionalism in a static field like architecture—by stating that an "object should look like what it can do." An object not only must remain faithful to materials, structure, and manufacturing processes used in construction, but it also must *express* its function when in use. As Geddes opined, "the intended use is the key to the form."[18]

This formula often yielded purely functionalist results. Geddes noted that a refrigerator's shape was "determined by the requirements of insulation and of storing various foods and containers of different sizes with proper regard for temperature, odor transference, and accessibility."[19] On the other hand, he described his Graham-Paige automobile (figure 37) with its "simplification of line" as "an artistic conception, based, of course, on the function it is to perform."[20] Successful design in this case depended on Geddes's aesthetic interpretation of the automobile as a fast vehicle. According to Geddes, a designer chose production methods and materials both "for their practical value within the object and for their external effect." To achieve an expressive effect he would permit "accents and embellishments for the sake of variety, contrast and emphasis" as long as they were para-doxically "achieved by use of the material functionally rather than in an applied manner."[21] Despite functionalist rhetoric, Geddes considered the crucial factor in visual design to be "the *idea*" behind it, "which is of an emotional nature."[22]

Most other designers advocated some form of functionalism modified by an element of expressionism. Franklin E. Brill of General Plastics referred to "a new functional language—a language of geometry, spheres, cubes, cones, squares, triangles" provided by machined molds—but he expected plastic products obtained from them to be "frankly expressive of the machine-age which made them possible." And Russel Wright, discussing radio design, asserted that by "stripping" radios "to their bare essentials, the modern designer invests them with a clean, fresh, beauty—functionalism at its best," but he also thought the designer should provide a "dramatization of inherent radio character." Even Teague described his Stormoguide barometer (figure 58), a square of Bakelite with an octagonal dial opening and chrome reeding across its sides and front, as "simple and practical but not too dully utilitarian."[23]

Other designers willingly went beyond functional design to expressive design. Peter Müller-Munk, director of an industrial design program at Carnegie Institute of Technology, taught students to perceive both "sensuous value inherent in pure form as produced by the machines" and "possibilities of creative expression through them." And Donald R. Dohner admitted that "while the intelligent and competent designer's forms may not always be exclusively functional, he will always design to express function." What a critic might perceive as an expressionist divorce of form and function reached its ultimate statement in Geddes's *Horizons.* Although he defined an object as "well designed" when it is "reduced to its utmost simplification in terms of function and form," he concluded that function, determined by an engineer, "is fixed," but "its expression in form," determined by an artist, "may vary endlessly under individual inflection."[24] Visual appearance, it seemed, was indeed superficial, related to function only by a designer's whims.

Geddes reached the knot of this functionalist-expressionist dilemma when he argued that modern design required two basic considerations. First, a design must be "created out of the elements, characteristics and needs of contemporary life," and second, it must remain true to "the Machine" as "the symbol of the age." Geddes's friends the Cheneys took a similar line in their volume on industrial design, *Art and the Machine*. Industrial design, according to them, required adherence to three principles, two functionalist, one expressionist. A designer should care about honesty of materials, simplicity of materials and forms, and "functional expressiveness"—the latter's purpose being "to bring out . . . a characteristic and expressive appearance."[25] Industrial products not only had to be functional. They had to look functional.

This crack in pure functionalist dogma eventually opened the way for complete commercialization of industrial design. Gradually, designers learned not to express personal perceptions of the machine age but to embody in their designs images of technological and material progress reflected back from the public. The function of attracting consumers would determine a product's form. But in the thirties few designers seemed hypocritical when discussing the machine style they were creating. Surprisingly, it was Teague, most business-oriented of the major designers, who rejected the expressionist interpretation of functionalism and opted instead for an uncompromising theoretical purism.

Teague sometimes fell to the machine's seductions. He was capable of finding "the spirit of this day" in such images as "a sleek dirigible against the sky, the soaring rectangularities of a skyscraper, the soft purring of an electric armature." But for a design philosophy he turned to the classical certainties he had learned from dynamic symmetry. He did not advocate imitation of classical forms. Instead he thought machine production had affinities with the impersonal, inorganic tone of classicism. The machine style's beauty might seem a novelty, but it was "familiar in Athens: the beauty of precision, of exact relationships, of rhythmical proportions; the beauty whose first law is perfect fitness and performance, whose second is candor and direct simplicity, whose third is a harmony of elements mathematically exact." Its "right lines, accurate and uncompromising angles, gleaming surfaces" signified "an integration based on necessity and not on caprice."[26] The precision necessary in machine production, in cutting molds and dies, revealed "the beauty of the exactly straight line, the perfect circle, the precise arc, the definite curve" and brought "today's machine art . . . closer in spirit to pure Classic than anything the world has ever seen since that day."[27]

Because the machine itself both required and supplied this inorganic precision, Teague thought that industrial design provided no outlet for artists bent on personal creation. An artist who sought "self-expression" should paint screens or carve bookends rather than enter industrial design.[28] Teague predicted that vagaries of individual inspiration would disappear—replaced by a universal "Machine Age aesthetic" relating all artifacts of civilization.[29] With or without conscious intervention by designers, a modern style would "create itself . . . out of the preferences and the prejudices of this age, out of these tools and materials and processes we use, out of the very functions which our products must perform." Rather than creating this style, Teague and his colleagues would derive it "by discovering the forms which fit our needs and our machines with the inevitability of a mathematical formula."[30] Its "unchangeable, invariable" principles lay embedded in "the structure of this universe and the structure of our perceiving minds."[31] With such a primal source, the machine style was "no ephemeral fashion, to be donned like the latest mode from Paris, discarded tomorrow." It was "here to stay."[32] Its products had a Platonic "finality of form."[33]

In idealistic moments Teague thought that "designs are always latent in the things" the designer "deals with, and it is his job to discover and reveal them."[34] Even when addressing the practical Society of Automotive Engineers, Teague insisted that a designer was "not an inventor trying to create new and unprecedented forms" but "an explorer seeking the one perfect form concealed within the object beneath his hand." True design meant discovering forms rather than applying motifs. Perfect form resulted from "perfect adaptation of means to end"—functionalism, in other words. "The design of any object, natural or man made," Teague asserted, "is inherent in the object itself, and must be evolved inevitably out of the function which the object is adapted to perform, the materials out of which it is made, and the methods by which it is made." Whatever the complexity or relative importance of a product, the goal is always "a perfect form." Teague recognized that this "ultimate form" might remain elusive. Since technological innovations followed one after another, in practice the perfect design for a given object would "be the last to be achieved before the object itself is obsoleted and superseded by some wholly new form." At a given moment, however, there exists a perfect solution to each problem, and the designer who discovered it would be rewarded with "beauty ... the visible evidence that design has been successful."[35] Since beauty possesses transparent obviousness, clients and public alike would recognize it and accept it.

For Teague, industrial design took on greater importance than as a sales device for products. Like Geddes and Giedion, he recognized that mass-produced goods shaped the everyday environment in which most people lived. With the introduction of mass-produced housing, industrial design would become still more significant. The human race could hope to live "in a setting of grace, charm and beauty" after its "envelopment by a machine-made world" only by expanding the industrial design perspective to include the entire built environment. "Conscious, intelligent planning," he wrote, "the analysis which sifts possibilities of variation for the one form functionally right, the kind of organization which evolves plastic unity out of complex assemblies of parts—industrial design, in short, offers the only hope that this mechanized world will be a fit place to live in."[36] Echoing Mumford's hopes for civilization's neotechnic phase, Teague predicted "a new world redesigned for human living," in which waste and ugliness would be replaced by efficiency, economy, and beauty. Blighted cities, gaudy commercial strips, and landscapes strewn with industrial debris now seemed "improper and antisocial." Usually a champion of individual rights, Teague found himself questioning "every man's inalienable right to make his own property as hideous as he pleases in his neighbour's eyes." But he expected "hard-headed industrialists," aware of the poor economics of waste, to lead the nation's aesthetic and social reconstruction.[37]

At his most optimistic he foresaw a new golden age—a society marked by harmony of spirit engendered by the machine, whose earlier disharmonies had culminated in the Depression. Teague agreed with Mumford that this new age would embody a degree of social stasis. Humanity would share an ideal society, given over—like the distribution machine of the consumer engineers—to endless repetition of frictionless process. A hint of the social control of totalitarianism appeared in Teague's vision of the future, but negative implications of a designed society hardly entered his mind, caught up as he was in projecting a harmonious future.

No other designer attained the coherence of Teague's thought, but some did propose applying industrial design to society. Geddes declared in *Horizons* that the new era would be "characterized by *design*." Not only would design eliminate factory drudgery and ease the lives of consumers, but, applied to "social structure," it would "insure the organization of people, work, wealth, leisure." He hoped eventually to see "a self-adjusting, economic mechanism" formed through "rational coöperation" of giant monopolies under government regulation.[38] More typical, however, was "Ten Years from Now," in *Ladies' Home Journal*, where Geddes predicted such wonders as double-decker streets, automated meal service in the home, holograms, synthetic paper, space exploration, and eradication of disease.[39]

Even more typical was a vision presented by Dreyfuss in *House Beautiful*. Directing attention to multiform changes occurring in "everyday surroundings," he listed some "of the better-looking things *within the reach of all*" in the Depression year of 1933. He pointed out the "new, good-looking refrigerator, sink, or stove" in "your kitchen"; furniture of "new materials and color combinations" in "your living-room and bedrooms"; "new ideas and conveniences in your bathrooms"; and the new auto "parked outside your door."[40] As long as industrial designers concentrated on pushing consumer goods, the goal shared by Teague and others of moving "toward the style of the century" would remain a pipe dream countered by a non-Platonic maxim: "Successful styling implies progressive restyling."[41] Nothing better demonstrated this conflict than the search for a perfect automobile.

The Perfect Aerodynamic Form

Readers of *Scientific American* must have received a jolt when they looked through the issue of January 11, 1913, and found an artist's conception of the "car of the future" —one of the first of many such visions (figure 91). Accustomed to the open touring car with surrey top and high-backed seats, they were amazed by this "long cigar-shaped

body" that looked like "a submarine boat more than . . . a carriage." Fully rounded in cross section, this hypothetical automobile had a bluntly rounded nose and tapering tail. The flow of its enclosed fenderless body— begun by a curved windshield, then technically impossible—was broken only by headlamps, running boards, rainwater moldings, and an exhaust pipe—all rendered unobtrusively. With such a design, according to the article, "wind resistance and the danger of skidding" would be "reduced to a minimum." After reviewing changes in body design from 1903 to 1913, the author concluded that "the tendency toward 'streamline' bodies is clearly evident, although opinions differ as to the ultimate design which will be evolved." He had no doubt that an ultimate body would be achieved, and that it would resemble somewhat the sketch he had presented.[42]

The idea of reducing wind resistance and boosting efficiency by means of a carefully shaped vehicle was not original. Isaac Newton had provided a basis for the search for the perfect vehicle in his law of fluid dynamics. According to him, the resistance of a fluid, whether liquid or gas, to a solid body varied directly in proportion to the square of its velocity, the square of its surface area, and the density of the fluid.[43] Obviously, a high-speed vehicle would require less energy to reach a given speed if its front surface presented as small an area as possible to the wind. Or so concluded one Medhurst, whose "New System of Inland Conveyance," a rail car proposed in 1827, would have been "tapered at both front and rear ends [to] move through still air or head winds with a minimum of resistance."[44] In 1865 a patent for a similar train design, never built, was granted to the Reverend Samuel R. Calthrop of Roxbury, Massachusetts. Thirty-five years later the Baltimore and Ohio tested its wedge-shaped *Windsplitter*, designed by Frederick Upham Adams.[45]

By the time auto-body designers began to think about wind resistance, more sophisti-

91. The car of the future, 1913. Scientific American.

cated aerodynamic theory had substituted the teardrop for the wedge as an ideal form. Late in the nineteenth century attention had focused on forces acting on the rear of an object headed into a wind or current. Scientists postulated a boundary layer of rolling tubes of air acting "much as roller bearings over which the outer air moves." Unless this boundary layer converged smoothly at the rear of an object, it would break into a chaos of retarding eddies. The ideal form would begin with a broadly rounded front, which would gently guide the boundary layer along a tapering body ending in a point. Since incorporation of wheels would make it impractical to round a vehicle's underside fully, engineers came to conceive of the ideal form for a land vehicle as a teardrop, the flat-bottomed shape taken by a drop of water as it glides with least resistance down a smooth surface.[46]

Before 1920 European body designers began applying this new interpretation, but its influence remained limited to custom-built racing cars. In 1911 W. G. Aston admitted in a British publication that the true attractions of good body design were "improved comfort and enhanced appearance." He argued nevertheless that "low wind resistance goes hand in hand with good appearance." Expanding on this statement, he claimed that "air behaves in certain circumstances much the same as the aesthetic sense behind our optic nerves, and it exhibits the same reluctance to adapt itself to bluff surfaces, sharp corners, ugly gaps, and ungraceful ... curves as does our vision—if properly trained." Thus "obtaining 'appearance'" and "reducing resistance" were for Aston "two targets which can be both transfixed with one shot." A perfectly streamlined automobile would appeal to an innate sense of fitness as the most beautiful form possible, but it would be necessary to educate the public to accept it.[47]

These ideas remained in the background until the late twenties, when aerodynamics became a growth industry in engineering departments of American universities. In 1926 the Daniel Guggenheim Fund for the Promotion of Aeronautics granted $2,500,000 for wind tunnels at the Massachusetts Institute of Technology, the California Institute of Technology, the University of Michigan, and New York University. Their research staffs concentrated on testing airplane forms, since "cleanness" was imperative with increased speeds of travel, but they also experimented with land vehicles like trains and automobiles. Theodore von Kármán, a Hungarian who came from Aachen to direct the California Institute of Technology laboratory, had tried unsuccessfully in 1924 to interest German auto manufacturers in body designs developed through wind-tunnel testing.[48] Most influential of Guggenheim engineers

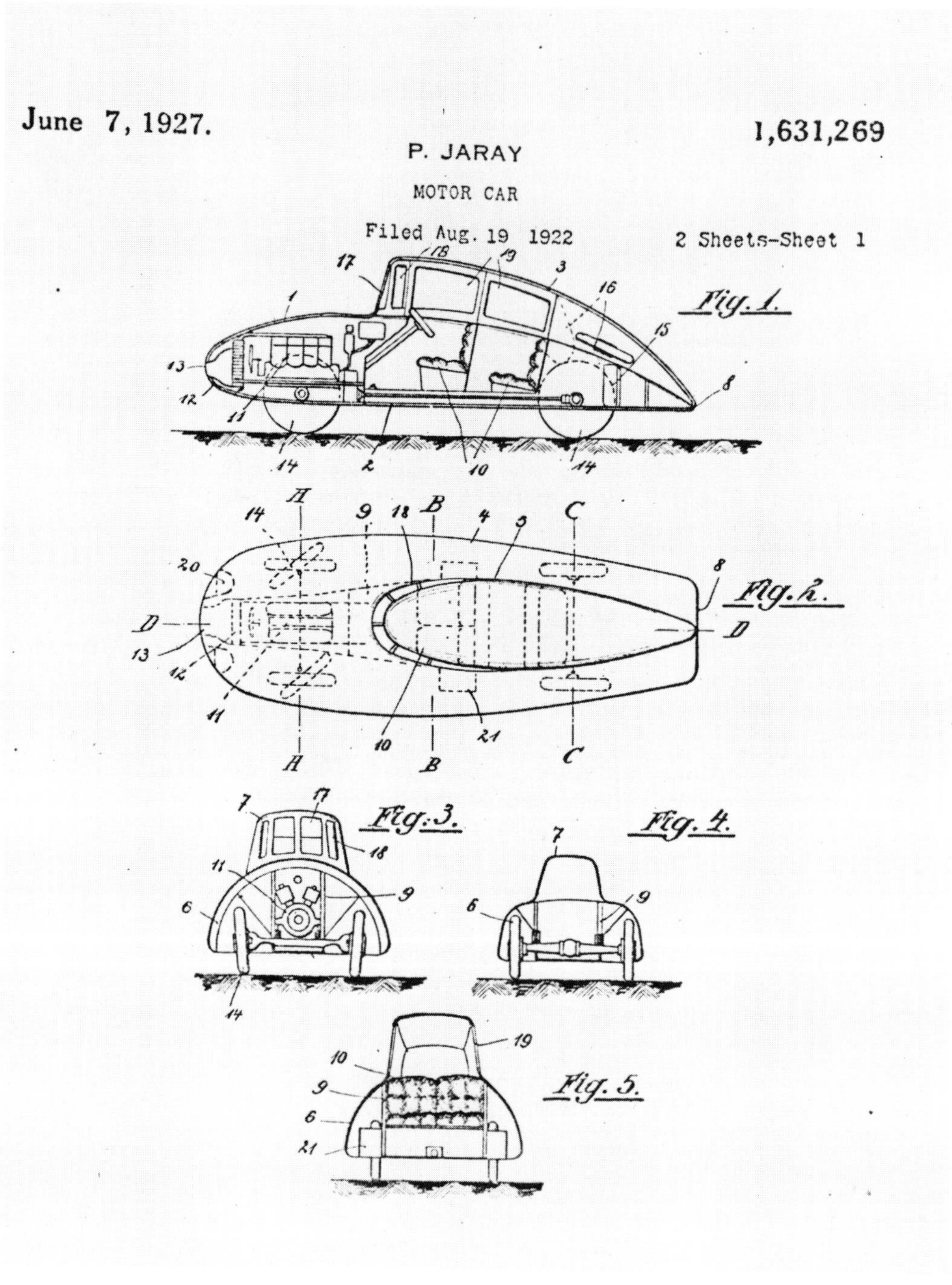

June 7, 1927.   1,631,269

P. JARAY

MOTOR CAR

Filed Aug. 19 1922   2 Sheets-Sheet 1

Fig. 1.

Fig. 2.

Fig. 3.   Fig. 4.

Fig. 5.

*92. Paul Jaray. U.S. patent sketch of partially
streamlined car designed in 1922. Geddes
Collection.*

engaged in automotive work was Alexander Klemin, who came to New York University from London. Probably in 1927 he tested a model of a partially teardropped car designed in 1922 by Paul Jaray, who had designed zeppelins before the war. His experiments brought the front-engine Jaray car (figure 92), flawed by a protruding hood, to the attention of American engineers and made Klemin an expert to be cited whenever the subject of streamlining autos came up, whether in technical journals or the popular press.[49]

Then in 1930 Detroit engineers learned of a revolutionary automobile designed in Britain by A. E. Palmer for Sir Dennistoun Burney (figure 93). By placing its engine in the rear, Palmer tapered the body completely without cramping passengers and provided a more fully streamlined front, although it bulged slightly below the windshield to provide a three-foot psychological buffer zone for passengers. A rear engine had other advantages as well. The Burney car had no running boards because eliminating the drive shaft made possible a lowered floor. Back-seat passengers enjoyed a smooth ride because they were seated between axles rather than over the rear axle. As *Automotive Industries* concluded in an enthusiastic review of the Burney car, it provided evidence of a "defined trend toward extraordinary streamlining."[50]

This ferment came to a head in Detroit during an annual meeting of the Society of Automotive Engineers in January 1931. By all accounts it was an exhilarating affair. The editor of *Automotive Industries* reported "an abundance of pungent engineering ideas darting about the sessions with something like the speed and power of electrons being ejected from a radio-active substance." Streamlining eclipsed all other subjects: "Possibilities for cars with engines in the rear and completely streamline or 'tear-drop' bodies" sparked "intensive thought in the sessions" and formed "the nucleus of more informal extra-session conversations than any other single idea generated." The teardrop form, he continued, "seemed to get almost unanimous approval as the final evolution" of the automobile body.[51] Among the "heavy guns" supporting the cause was H. Ledyard Towle, an ad agency art director and former color expert for General Motors, who opened the meeting with his address "Projecting the Automobile into the Future." According to Towle, the teardrop auto would, as "the perfect aerodynamic form," efficiently boost average speeds by twenty

93. A. E. Palmer. Rear-engine teardrop car designed for Sir Dennistoun Burney, c. 1930. *Chilton's Automotive Industries.*

miles per hour. Although the industry would have to manipulate public acceptance of such a radical form, inhabitants of the future would look back at the thirties as an "American Renaissance," when beauty and function converged perfectly. As if anyone had missed the point, H. T. Woolson of Chrysler emphasized in a discussion period that "the design that is technically correct is automatically good looking." The teardrop "might not at first appeal to the public, but, if it is fundamentally correct, it will grow in favor and be acceptable."[52]

During the next two years countless engineers searched for the perfect embodiment of the ideal teardrop form, which was accepted by virtually all. Some sought theoretical results. A professor of mechanical engineering at the University of Michigan, W. E. Lay, ran a variety of models through his wind tunnel, as did R. H. Heald of the National Bureau of Standards.[53] Others sought the perfect car itself. Both Walter T. Fishleigh, a Detroit consulting engineer, and O. G. Tietjens, director of aerodynamic research at Westinghouse, developed wind-tunnel models of rear-engine teardrop cars, which they expected to be aerodynamically perfect and technically feasible.[54] Most engineers assumed that production autos would exhibit perfect form as consumers became accustomed to it through ever closer approximations. As yet, however, interest remained limited to engineers. Auto executives considered production of a teardrop car an extreme move, both because it would require expensive retooling and because they were uncertain the public would accept it. But showman Norman Bel Geddes convinced one of them he might be wrong.

Geddes became interested in streamlining in the late twenties. Through the sport of sailing he discovered the "fascination of water—currents—eddies—waves—slipstream—streamlining."[55] Le Corbusier's *Towards a New Architecture* contributed to his growing interest; his copy, received shortly after its English translation in 1927,

is well thumbed. Le Corbusier compared evolution of the automobile form to the development of Greek architecture, culminating in the Parthenon's perfection. He discussed wind resistance and put in his book a chart that provided resistance coefficients of several forms, including the efficient teardrop (figure 94). From him Geddes learned that the teardrop was the ultimate automobile, which, once perfected, would remain unchanged.[56] An opportunity to develop streamlining came with the stock market crash, which left his office for several years with few commissions. To keep staffers occupied, he set them to work on streamlining. Not merely a blue-sky designer, Geddes provided some technical basis for their designs by conducting crude tests—first by dragging various wooden forms behind an outrigger attached to his sailboat and later by blowing smoke past them in a small wind tunnel on the roof of his office building.[57] The results appeared in *Horizons* in November 1932.

Ignoring the usual opinion of engineers that streamlining's goal was equal speed with less fuel, Geddes insisted that "speed is the cry of our era, and greater speed one of the goals of to-morrow."[58] Without analyzing the need for greater speed, he launched into a layman's exposition of aerodynamics followed by illustrations and descriptions of transportation designs developed in his office. Photographs of his streamlined train model probably influenced the design of the Union Pacific's first streamliner. A model of an ocean liner (figure 95) exhibited a tapering torpedo shape with two teardrop stacks and a wing-shaped bridge. The most astonishing design was a mammoth flying wing with a spread of 528 feet and teardrop pontoons (figure 96). Powered by twenty engines, it would have carried 450 passengers from Chicago to England in forty-two hours. For automotive engineers his designs for an automobile and a bus held the most interest. Motor Car Number 8 (figure 97) uncompromisingly embodied the teardrop form by placing the driver right behind its broad

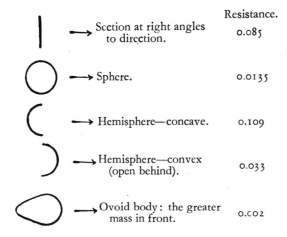

146   TOWARDS A NEW ARCHITECTURE

		Resistance.
|	→ Section at right angles to direction.	0.085
○	→ Sphere.	0.0135
(	→ Hemisphere—concave.	0.109
)	→ Hemisphere—convex (open behind).	0.033
⬭	→ Ovoid body : the greater mass in front.	0.co2

The cone which gives the best penetration is the result of experiment and calculation, and this is confirmed by natural creations such as fishes, birds, etc. Experimental application : the dirigible, racing car.

94. Le Corbusier. Air resistance of various forms, 1927. Towards a New Architecture.

front, psychologically unprotected. By extending the body over the front wheels, Geddes eliminated front fenders, but the car had two long, sweeping rear fenders, necessary for enclosing wheels while the body tapered between them. A rear stabilizing fin contained a gas tank—certainly an unsafe arrangement. The car would have accommodated eight passengers in a two-three-three seating pattern before tapering over luggage compartment and rear engine. His Motor Coach Number 2 (figure 98) echoed the design of Motor Car Number 8. *Horizons* also included illustrations and descriptions of scales and stoves, the Toledo Scale factory, the rotating airport, and a streamlined house. A vague conclusion exhorted readers "to cut loose and *do the unexpected*" by applying industrial design "to achieve that complete mastery of the machine which is to-day a more or less unconscious goal."[59]

Serious critics roasted the book as super-

95. Geddes. Model of proposed streamlined ocean liner, 1932. Geddes Collection.

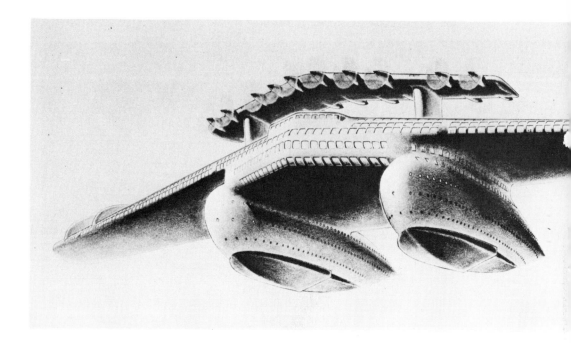

*97. Geddes. Model of proposed Motor Car
Number 8, 1932. Geddes Collection.*

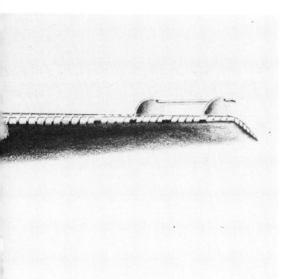

*96. Geddes. Sketch of proposed flying wing, 1932.
Geddes Collection.*

*98. Geddes. Model of proposed Motor Coach
Number 2, 1932. Geddes Collection.*

ficial and derivative. "As an exhibition of fresh thinking," wrote Lewis Mumford, *Horizons* was "comparable to the second discovery of America in the year 1592." Frank Lloyd Wright concluded that Geddes's "treatise on popular mechanics" had "little of value to offer either art or industry except as it becomes a sort of stage scenery continued for sensational effect." The most pointed critique came from Douglas Haskell, one of the decade's few architecture critics. Although he admitted finding "hints of principles at work," Haskell dismissed the book as "a lucid story," streamlined perhaps, "in which the technical details have been either flattened out or spirited away to form no obstacle to the untechnical reader." *Horizons* embodied "a double appeal—the ingenuities of popular mechanics are related to those of advertising psychology." Haskell concluded that Geddes addressed it "not to the discriminating individual" but "to the manufacturer and to the mass market."[60]

These criticisms contained a measure of truth. Even Van Doren, more sympathetic as a fellow designer, in 1940 described Geddes as "a genius of the theater, who put his talent and enthusiasm into the visualization of a world of the future, in which everything that moved through water or air or under the sea was somehow manipulated into the shape of a catfish." Streamlining, Van Doren recalled, "was in the air," and "what Geddes did was to dramatize it, well before it had really arrived, and so convincingly as to crystallize the scattered forces already tending in that direction."[61]

Horizons certainly succeeded as publicity for streamlining. The public learned of Geddes's futuristic designs through the Sunday supplements, but the book also provided automotive engineers with ammunition for convincing wary executives that the aerodynamically streamlined automobile possessed commercial potential. Carl Otto recalled that *Horizons* "had great impact" among his colleagues in the Art and Color Section at General Motors.[62] But Geddes's

book had greater impact at the offices of Chrysler, where it tipped the scale in the decision to produce the Airflow of 1934, the first production auto seriously to embody streamlining. Since 1927, originally assisted by Orville Wright, the company's engineers had been conducting wind tunnel tests and building prototypes (see figure 99), but despite eventual support from Walter P. Chrysler, business managers and marketing experts consistently rejected the project. A copy of *Horizons* sent by Chrysler himself to the "selling side" finally won approval. Three months before the Airflow was introduced to the public, Fred M. Zeder, head of the engineering department, admitted privately that *Horizons* was "entirely responsible" for giving the firm courage to proceed with it.[63]

Zeder and his associates intended the Airflow to have a pristine teardrop shape, but Chrysler's available engines made a rear-engine car "tail-heavy."[64] As finally produced (figure 100), the front-engine Airflow had a broad flat nose that curved into the hood. Headlights were incorporated with the body between integrated fenders and the radiator, whose horizontal chrome stripes extended beyond it, ending in an inverted V to emphasize the front end's continuity. The car's roof line was a parabolic curve ending in a sharply tapered rear enclosing the trunk. Teardropped rear fenders, equipped with removable wheel covers, flowed back smoothly. With the exception of fenders and narrow running boards, the Airflow had flat sides, but its hood and roof lines flowed in continuous curves. To taper the rear without cramping passengers, Chrysler engineers placed both seats between axles and obtained a smoother ride than ever before. A "functionalist" interior had tubular seat frames and chevron moldings on door panels. Although engineers predicted that an Airflow sedan driven at sixty miles per hour would cut only 15 percent of the wind resistance encountered by a standard sedan, in appearance it was radical. Alexander Klemin described the car

99. Prototype of Chrysler Airflow in wind tunnel, c. 1933. Arens Collection.

as a compromise "between ideal aerodynamics and practical automobile design."[65] Automotive engineers took heart at seeing a manufacturer introduce such a mutation into the evolution of the perfect automobile.

In order to consolidate this gain, Chrysler had to sell the public on the Airflow. Four months before introducing the car at the annual auto show, Walter Chrysler hired Geddes both to design automobiles and to coordinate publicity for the Airflow. Geddes arranged for a small group of celebrities and designers to sing the praises of the Airflow in magazine advertisements run shortly before the auto show. One featured Geddes himself, shown alighting from the car, which was not fully illustrated in order to build suspense for its public unveiling. The ad billed Geddes as "America's foremost industrial designer" and author of "the authoritative treatise on modern design," which had "forecast Airflow motor cars." Geddes himself claimed the Airflow as "the first sincere and authentic streamlined car . . . the first *real* motor car." He went on to state that "when the line and contour of a motor car start with the problem of what a car is designed for . . . and when the outward design and interior engineering combine in an harmonious structural whole . . . the result is a great motor car!"[66] Initially, the public agreed. When the Airflow appeared on center stage in New York in January 1934, it drew record orders.

100. Chrysler Airflow Imperial coupe, 1934. Courtesy of Automobile Quarterly Magazine.

Geddes, however, thought privately that it did not go far enough. During his brief association with Chrysler, he tried to move the company beyond the Airflow to the perfect teardrop. Geddes's assignments included a superficial restyling of the Airflow's front end to be incorporated in mid-season. But from the start he stressed development of a fully streamlined car. At one of his first meetings with Walter Chrysler, in October 1933, Geddes presented blueprints of four automobiles designed by his staff. At least two were teardrops. Insisting that such cars "wouldn't be seen on the roads for the next five years in any kind of production" because "the public is not yet educated to streamlining," Chrysler offered Geddes five dollars "even money" to back up his opinion.[67] By the middle of November Geddes seemed on his way to winning the bet. Roger Nowland, his aerodynamics expert, presented Chrysler and two top engineers with refutations of their objections to rear-engine cars. If front ends were resilient instead of rigid, they would provide collision protection. Proper location of the center of gravity would eliminate "tail-sway" and rough riding— both common in experimental rear-engine cars. As for the argument that a rear engine might be thrown forward in a crash, Now-land rejected it out of hand. Geddes and his men left Chrysler with tentative agreement to the production of a rear-engine car.[68]

A meeting in January 1934, however, found the tireless designer arguing with Chrysler engineer Carl Breer that the company should take advantage of its "two year jump on the industry" by introducing a rear-engine teardrop car. Breer countered by stating that Chrysler, who was not present, would not even consider Geddes's "present conception of ideal car design from an aerodynamic and aesthetic viewpoint" until he provided "concrete recommendations in restyle of the present line." But Chrysler's engineer yearned like Geddes for "the perfect aerodynamic form." On his own authority Breer left Geddes with an agenda specifying not only a "1934-1/2" Airflow restyling but also a design for 1935 or 1936 embodying "ideal streamlining, letting engine space come where it may."[69] A week later, however, Chrysler himself told Geddes to downplay the project. The stubborn designer left one man working on it but concentrated his staff's efforts on the Airflow restyling. Geddes's designs reached only the quarter-size model stage of development. He failed even to convince Chrysler not to provide the 1935 Airflow with a V-shaped radiator.[70]

Thus ended their tug-of-war on the question of the perfect automobile.

If doubts lingered in Chrysler's mind, they faded as the Airflow's sales performance became known. The company failed to reach full production of the car until April 1934, long after official introduction. Many orders evaporated. To make matters worse, the first two thousand off the line exhibited serious production defects. The Airflow gained a reputation as a lemon. In 1934 about 8,500 were sold. By comparison Chrysler sold three times as many conventional six-cylinder sedans. As a result, instead of making successive Airflows more radical in appearance, Chrysler introduced increasingly conventional designs until 1937, when he dropped the car from the line. Donald J. Bush has suggested that the Airflow failed because it lacked the teardrop shape associated by the public with perfect streamlining.[71] If so, it failed because it was not radical enough. Whatever the cause, its failure guaranteed that Chrysler would proceed no further toward the aerodynamically perfect automobile. Other manufacturers who had experienced "great soul struggles" when the Airflow appeared now also rejected radically streamlined designs.[72]

Writing of the car of the future in March 1934, Loewy said that "there will be nothing to disturb the smoothness of its tear-drop silhouette as it cuts through the air," but automotive engineers had begun to doubt that the teardrop did represent the ideal form.[73] Gathering in the Adirondacks three months later for their summer meeting, they addressed just that question. It was, as *Automotive Industries* reported, "one of the most widely discussed topics."[74] William B. Stout, among the first to campaign for the teardrop twenty years earlier, claimed that wind-tunnel results proved nothing because they did not measure effects of side winds and of the ground itself. The ideal car would have sloping sides and a smooth bottom. Ironically, it would resemble a turtle.[75] Herbert Chase, who had championed the rear-engine

teardrop three years earlier, now said of automotive aerodynamics that "nobody knows anything about it."[76] Others echoed this opinion. Although they had championed the teardrop with utopian fervor, most now would have agreed with architect Paul Philippe Cret, who had found while engaged in train design that "streamlining was still a hazy subject."[77] Even the notion that perfect streamlining would coincide with ideal beauty came under the engineers' attack. After Teague presented his Platonic design philosophy in his address "On the Artistic Principles of Body Design," Walter T. Fishleigh countered by arguing that "functionalism" did not yield "an obviously beautiful result" because the aesthetic sense is subjective and personal.[78]

With their own engineers uncertain about the form of the ultimate car, manufacturers quickly abandoned a quest that would hardly increase sales in the long run. General Motors, the industry's leader, adopted different positions on streamlining, one for the market, the other for insiders. Two pages in the company's 1935 *Automobile Buyer's Guide*, "Aerodynamics and Streamlining," declared that "simplified contours and a minimum of protruding parts,—the airplane type V-shaped windshield, the enclosed spare tire and the rear ends scientifically tailed out to reduce the drag caused by vacuum . . . add much to the beauty of the new General Motors cars." Nowhere did the company claim that a scientifically designed rear end would cut wind resistance, but car buyers were left to make that conclusion.[79] In a letter to stockholders, however, president Sloan noted that GM's 1935 models would reflect "the trend toward streamlining —a vogue very much in evidence at this time"—despite the fact that "the contribution of streamlining is definitely limited to the question of styling." Belief that streamlining would reduce operating costs or increase efficiency reflected "popular misunderstanding."[80]

Sloan was willing to take advantage of the

101. Loewy. Rendering of Princess Anne *for Virginia Ferry Co., 1936. Raymond Loewy.*

public's ignorance. As long as people demanded streamlining, General Motors would provide it in annual dollops. Early in 1935 Charles F. Kettering, director of research and development at GM, defended annual model changes because they embodied technological improvements—a claim countered by Sloan's 1926 rationale for the Art and Color Section. "We have not," Kettering went on, "*depreciated* these old cars, we have *appreciated* your mind" by providing you with "something better than you knew there was in existence."[81] There was no end in sight. Each year would bring the new improved "practical ultimate."[82] As manufacturers competed for customers, automobiles would exhibit a "search for individuality and differentiation" characteristic of "designing for eye resistances rather than wind resis-

tances."[83] The engineers' search for the perfect form dissolved, replaced by a quest for sales. The teardrop survived only in Flash Gordon's futuristic frames and in popular articles on the "car of the future"—there serving as stimulus for consumption of less perfectly streamlined goods and services. Even Loewy was "the first to admit" that he streamlined the superstructure of the *Princess Anne* ferry (figure 101) not to reduce wind resistance but "to produce a design that embodied something different and daring which would appeal to the taste and attention of the traveling public, already quite familiar with the applications of artistic air-flow design in rail and motor transportation."[84] Once a part of aerodynamics, streamlining became a matter of styling.

From Depression to Expression

Simple lines . . . tend to cover up the complexity of the machine age. If they do not do this, they at least divert our attention and allow us to feel ourselves master of the machine.

—Paul T. Frankl

Visitors to the 1933 Century of Progress Exposition in Chicago thought they glimpsed the future they would confront once the heat of accelerating technologies had burned off Depression fog. Although many exhibits looked back at the changes made since Chicago had emerged as a frontier settlement, more looked forward to the future. Wonders like the photoelectric cell promised a life of ease, and assembly lines displayed by Firestone and General Motors testified to industry's undiminished technological clout. The architecture itself, stretched around a lagoon and along Lake Michigan, awed most visitors. Conceived by the architects who had recently provided New York with its modernistic skyscrapers, the fair's buildings were fantastic extensions of the same style. Jagged setback towers, monolithic pylons, and rugged vertical fluting dominated the structures. Most appeared machined with precision tools by a race of giants. The four huge angular blocks of the General Exhibits Group (figure 102), with identical asymmetric towers, looked as though they had come off an assembly line themselves.

Architects had distinguished the composite forms of the fair's massive angular buildings with large areas of brilliant colors, most commonly blue, gold, orange, and deep violet. Street vendors capitalized on this kaleidoscopic scene by selling tinted glasses to "protect [the] eyes against the sun and the buildings." Douglas Haskell complained that "what had not already been fretted away in the jagged forms was frazzled by the disparate hues." The fair's architecture produced a total effect "more curious than beautiful, and everywhere restless."[1] Purged of the stylized organicism and precious details of the Paris exposition, the modernistic style had yielded completely to machine-age rhythms.

Most informed observers agreed that the fair's architecture would have a permanent impact, whether for better or worse: Harvey Wiley Corbett of the architectural commission described the fair "as a symbol of the

102. General Exhibits Group, Chicago Century of Progress Exposition, 1933–1934. Official Pictures of the 1934 World's Fair.

architecture of the future."[2] Few realized that the modernistic style, a commercial creature, was doomed to a short life. Its upbeat rhythms harmonized with the twenties, but the style did not satisfy Depression needs. Economic collapse brought enough dislocation. People did not need artificial stimulation. They needed reassurance that progress would be fast but smooth, not feverish or jerky. Above all, they sought security. Since life seemed too complex to comprehend, people gravitated toward objects whose surfaces seemed smooth, simple, and efficient. As architects and designers perceived this shift in mood, began to intuit and objectify unconscious predilections of thousands of individuals, they created a new style —a new packaging for products and commercial architecture of the Depression. Like the modernistic style, this new mode of expression embodied technology as the central fact of modern life. The modernism that cul-

minated in the Century of Progress Exposition had taken inspiration from the staccato rhythm of the mass production assembly line, but the new style owed its forms to transportation machines whose effortless passage reflected the ease with which most people wished they could glide through the Depression. Not a creation of intellectual machine-age aestheticians, the streamlined style drew on visible popular examples of technological innovation.

See America Streamlined

At the exposition itself a shadow of the style that would render its own obsolete was cast by a cigar-shaped Goodyear blimp floating above the grounds. Present in backgrounds of many publicity photos, the blimp reveals its significance to us clearly in a colored postcard illustration of the General

Motors building (figure 103). A sky of thick blue-black clouds, filling three-quarters of the picture, announced an impending storm. Into this dark atmosphere GM's brilliantly lit orange tower projected confident defiance, while tourists hurried below in search of diversion, as if protected from the storm by its glare. Over everything floated the Goodyear blimp—cool, distant, a dull metallic gray, capable of riding out the storm. Its smooth unbroken skin contrasted strikingly with the hyperactive people and angled architectural forms below. From its perspective they dwindled to insignificance before the storm, while from below the blimp looked like a command post, projecting an image of serene control. The fair's modernistic architecture seemed like whistling in the dark. But the blimp—its shape aerodynamically determined for greatest ease of passage through air—suggested the idea of streamlining society itself to bring it through Depression turbulence.

The Goodyear blimp probably did not produce such reflections in fair-goers who watched it. Drifting aloof, exemplifying elitist transportation, it lacked immediacy. Oddly, a traditional transportation form—the railroad—made streamlining a popular fad and propelled its transformation into a design style. During the exposition's first year two major roads, the Union Pacific and the Chicago, Burlington and Quincy, planned and constructed streamlined passenger trains. Introduced in time for display during the fair's second year, these two streamliners became "symbols of an era." As *Business Week* concluded in 1940, when the UP's *M–10,000* and the Burlington *Zephyr* "made their bow at the Chicago Century of Progress, something clicked in the mind of Mr. Average Man, and the streamlined era had arrived." Thanks to their influence, streamlining became "a merchandising asset for industry at large." [3]

The Depression hit railroads harder than many segments of the economy. Since 1920 they had been losing passengers to private

automobiles, bus lines, and eventually airlines. By 1930 autos had captured three-fourths of the traffic between cities, while buses enjoyed a fifth of the remainder. A general increase in volume of travel somewhat softened rail losses, but the annual number of passenger miles traveled by rail fell from a high of 47 billion in 1920 to a low of 16 billion in 1933—the lowest total since 1900. Most companies supported passenger operations with freight revenues, but these evaporated during the Depression. Somehow passenger operations had to be made self-supporting.[4]

Industrial designers weighed in with advice. Egmont Arens sent reprints of his 1931 article "The Train of Tomorrow" to executives of every major railroad. He felt certain that increased speed would bring passengers flocking back. And new technologies, he argued, would permit increased speed with considerable savings. Cars of new aluminum alloys would be light enough to be pulled by a diesel engine—a motive power more efficient and less costly to fuel than the steam engine. Wind resistance could be reduced and speeds of up to 120 miles per hour attained, he claimed, by putting the cab in front of a bulbous locomotive, tapering the observation car like a fish's tail, lowering a train's center of gravity, and bringing sealed windows flush with a smooth exterior.[5]

Arens's publicity effort whetted executive appetites, but they went unsatisfied until Geddes's publication of *Horizons*. Geddes probably realized when he introduced his own streamliner that current four-cycle diesel engines were too heavy for mainline railroad operation. His proposed steam train would bear no resemblance to predecessors. A smooth aluminum shell would enclose smokestack, pipes, boiler, and headlight. Metal shutters would cover driving wheels and trucks. A smooth elastic material would cover openings between cars. The train would become a single tube with gently rounded sides and roof, a gradually bulging

103. Postcard showing General Motors building
and Goodyear blimp, Chicago, 1933. David T.
Meikle.

front, and a tapered rear. As illustrated in *Horizons*, its gray shell highlighted against a black background (figure 104), Geddes's train seemed the ultimate in fast, efficient travel of the future. Soon after publication of his design, both the Union Pacific and the Burlington, each under new management, began planning lightweight streamliners. Although the degree of his influence is uncertain, contemporary observers noted the UP train's likeness to Geddes's design.[6]

The UP and the Burlington reached similar solutions in their first streamliners. Each train consisted of three cars: two day coaches and a locomotive with baggage section and mailroom. On each train the joint between cars was articulated, meaning that the ends of both cars rested on and pivoted on a single truck of wheels. This coupling system eliminated the weight of two trucks, made a train more stable on high-speed curves, and gave it the appearance of an unbroken tube.[7]

To avoid the conservatism of railroad men, both companies hired structural engineers with experience in newer transportation industries to develop lightweight cars. Edward G. Budd, who had pioneered all-steel welded auto bodies, provided the Burlington *Zephyr* with stainless steel cars welded in a new electric process that joined stainless steel without weakening it or destroying its rust resistance. And William B. Stout, who had built the first all-metal airplane, transferred his aviation experience to the UP train, whose cars boasted a load-bearing tubular frame of extruded Duralumin covered with sheets of the same alloy. This monocoque construction eliminated the heavy steel underframe beam of traditional rail coaches and contributed to the train's aerodynamic streamlining, which was perfected at the University of Michigan. The Burlington train received its aerodynamic imprimatur from the Massachusetts Institute of Technology.

For generating power, the UP relied on a 600-horsepower V–12 engine, but Ralph

104. Geddes. Model of observation car of proposed streamlined train, 1932. Geddes Collection.

Budd, president of the Burlington, insisted on a diesel. Entranced by two experimental two-cycle diesel engines made of new lightweight steel alloys, which powered the assembly line at GM's Chicago fair exhibit, Budd pressed Charles F. Kettering for an engine. Completed ahead of schedule, GM's 600-horsepower diesel took its place in the *Zephyr*. Lightweight, efficient, and designed to cut wind resistance, these new trains were streamlined in materials, methods of construction, and motive power. Each three-car train weighed roughly the same as a single Pullman car, while wind-tunnel tests suggested that at maximum speed they would halve the amount of energy required by a steam locomotive with two standard coaches for overcoming wind resistance. Despite innumerable technological innovations, however, the public responded to appearance— and to the promise of speed indicated by appearances.[8]

The UP's *M–10,000* (figure 105), later renamed *The City of Salina*, was completed in February 1934. *News-Week* described it as "a great bulbous-headed caterpillar," but it looked like a snake as it whipped along the track.[9] The continuous horizontal flow of the train, created by articulated couplings

and the low, rounded cross-section of the cars, received emphasis in the exterior color scheme. Along each side swept a wide bright yellow band, terminating near each end in a half circle. This band provided an electrifying contrast to the rich brown of the rest of the train. The rear car tapered slightly to end in a half sphere unbroken by windows and faired under to conceal the rear truck. The locomotive had two discrete elements —a semicircular air scoop gaping like a shark's open mouth and, perched above it, the cab's rounded frame, surmounted by a projecting headlight. Although the front end looked ferocious, the overall impression was of smooth forward motion.

The Burlington *Zephyr* (figure 106) was completed in April 1934 and later renamed the *Pioneer Zephyr* to distinguish it from succeeding Burlington streamliners. Its exterior remained unpainted because the stainless steel's color "expresses the motif of speed."[10] It did not present as clean an appearance as the *M–10,000*. Side panels, horizontally corrugated for strength, appeared functional and enhanced horizontal flow but violated an obsession with smooth surfaces soon typical of the aesthetic of streamlining. Handrails by each door and

105. Union Pacific's M–10,000, *1934.* Travel.

106. Burlington Zephyr, *1934. Hawthorn Books, Inc.*

sheathing around wheel trucks interrupted smooth visual flow. The truck sheathing might have limited retarding eddies of air created by turning wheels, but it looked tacked on because it did not extend the length of the train. Less pretentious than the front of the UP, that of the *Zephyr* in profile looked like a straight line slashing down at a slight angle. While the UP train evoked comparisons with animate creatures, the *Zephyr* looked like a machine, pure and unvarnished. Even its interior of Flexwood, Masonite, and chrome was intended by architect Paul Philippe Cret as "an ensemble where modern fabrics, synthetic materials, would enhance the stainless steel of the framework."[11]

These first streamliners bore slight resemblance to each other, but they both departed from all previous trains. These "wingless airplanes on tracks," as Stout called them, evoked pandemonium wherever they went.[12] Schools let out, brass bands played, mothers held up babies for a look at the future, and crowds gathered at rural crossings to witness exhibition runs. Fifteen million people glimpsed one train or the other in the summer of 1934. On tour for twelve weeks, the *M–10,000* traveled 13,000 miles on the tracks of fourteen railroads in twenty-two states and made sixty-eight stops to allow people to go through the cars. The *Zephyr* covered 30,000 miles and stopped at least briefly in 222 cities. In a single day in Philadelphia 24,000 visitors went through the *Zephyr*. A two-day stop in Chicago drew 34,000. During a three-week tour of the east coast, 380,000 people trooped through the train, while a two-week sweep of the Midwest netted 105,000. On May 26, 1934, the opening day of the Century of Progress's second season, the *Zephyr* highballed from Denver to Chicago in fourteen hours, cutting twelve hours from the time of scheduled trains and arriving in time to climax a performance of "Wings of a Century," a pageant dramatizing the history of transportation. Throughout the summer, 709,000 fair-goers patiently waited in lines to go through the

Zephyr (figure 107). A similar number toured the *M–10,000*, which soon joined its competitor.[13]

The railroads played their new streamliners for publicity, but the public took to them without prodding. Other streamlined vehicles had recently appeared. At the fair, Buckminster Fuller's Dymaxion Car No. 3 (figure 108), a three-wheeled, rear-engine teardrop, played in "Wings of a Century." And Chrysler had just introduced the Airflow. But the trains crystallized interest in streamlining. Highly visible, they represented a bit of the future available to anyone who could afford a ticket. Within two years a dozen railroads responded to public exuberance with orders for thirty more streamliners, both diesel-electric and steam, and more followed. Most diesel locomotives resembled the *Zephyr* in profile but had smoother lines, usually accented by use of colors. Some "steamliners" followed the *Zephyr* model, but most exhibited bullet-nose forms originally devised by Otto Kuhler (figure 109) and popularized by Loewy for the Pennsylvania. Often companies refurbished older locomotives with tons of sheathing, enclosing boiler, pipes, and stack in a smooth shell. Such redesigned locomotives attracted passengers, but any gain in aerodynamic efficiency was offset by added weight. The intention was commercial. In fact, true aerodynamic elements like articulated coupling and sheathed trucks yielded to more traditional—and more practical— arrangements.

Dreyfuss, Loewy, Teague, and a host of specialized designers and engineers benefited from the rapid acceptance of rail streamlining. Each railroad vied for passengers with the most novel-looking equipment it could muster. The years from 1934 to 1939 constituted a "craft" phase of streamlining because each company operated custom-designed equipment. But mass production won out. The Electro-Motive Division of General Motors in 1935 began constructing a five-million-dollar plant for

107. Burlington Zephyr *on display at the
Century of Progress Exposition, 1934.* Official
Pictures of the 1934 World's Fair.

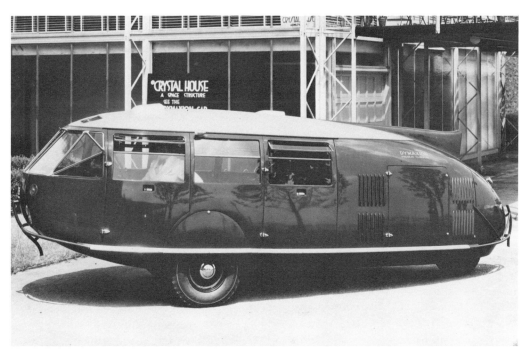

*108. Buckminster Fuller. Dymaxion Car No. 3,
1934. Buckminster Fuller.*

109. Typical steam locomotive with streamlined sheathing suggested by Otto Kuhler, 1931.
Railway Age.

manufacturing complete diesel locomotives. By 1939 GM had delivered seventy diesels with uniform bodies, distinguished only by engine specifications and individual color schemes for each railroad. Just before Pearl Harbor, GM was producing forty-eight loco- motives a month, including more powerful units for hauling freight. The heyday of the streamliner was over.[14]

In the meantime, however, streamlined trains stimulated public faith in a future fueled by technological innovation. As a publicist asserted, they provided "a portent of change and progress, and the first drama- tized portent of the kind since the crash of prosperity five years ago."[15] Businessmen moved quickly to profit from and take con- trol of this genuinely popular fad. Immedi- ately, streamlining affected such ephemeral cultural forms as movies, toys, novelties, and advertising. *The Silver Streak*, a film released by RKO late in 1934, starred the *Zephyr* itself, racing against time to deliver an iron lung from Chicago to the Boulder Dam. The following year brought *Streamline Express* as well as a Paramount film about a swing band touring in a monstrously streamlined bus (figure 110).[16] *Fortune* described streamlining as "the biggest thing that has

happened to the toy-train industry for decades." Both Lionel's *M–10,000* and American Flyer's *Zephyr* became bestsellers at twice the price of a steam engine and whipped around basements at 200 scale miles an hour. For the American National Company of Toledo, Van Doren and Rideout designed the Skippy Airflow, a pedal car followed by streamlined sleds, wagons, scooters, and tricycles. A typical trike (figure 111) boasted teardrop fenders, a broad curvilinear body of stamped steel, and webbed handlebars evocative of airplane wings. For playful adults the Streamline Lighter Company provided a chrome-plated egg-shaped lighter, which split apart to re- veal its striking mechanism.[17]

As a vague concept, streamlining became attached to almost everything. Housewives, who might already have purchased Geddes's "streamline stove" (figure 61), were ex- horted by a *New York Sunday News* colum- nist to "streamline the silhouette" with back- straightening exercises.[18] They could also grace their tables with "Royal Streamline" flatware, a spare but uninspired pattern.[19] Their husbands could purchase Firestone's streamlined tire, designed by Geddes late in 1934 with a series of flowing concentric

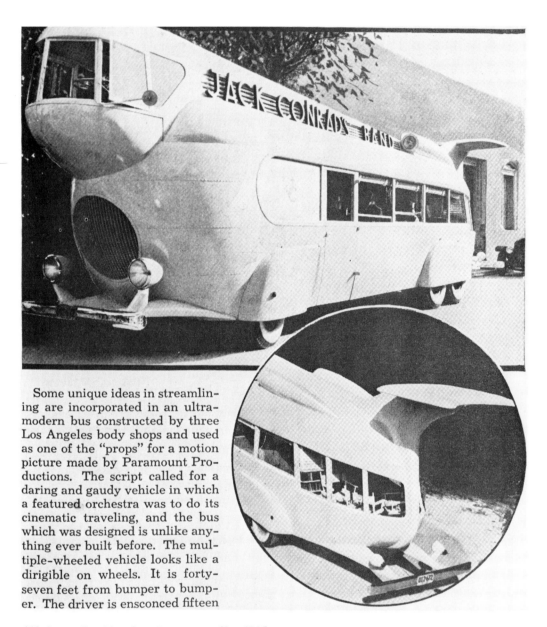

Some unique ideas in streamlining are incorporated in an ultramodern bus constructed by three Los Angeles body shops and used as one of the "props" for a motion picture made by Paramount Productions. The script called for a daring and gaudy vehicle in which a featured orchestra was to do its cinematic traveling, and the bus which was designed is unlike anything ever built before. The multiple-wheeled vehicle looks like a dirigible on wheels. It is forty-seven feet from bumper to bumper. The driver is ensconced fifteen

110. Streamlined bus for a Paramount film, 1935.
Popular Mechanics.

*111. Harold Van Doren and John Gordon
Rideout. Tricycle for American National Co., c.
1935. From* Industrial Design *by Harold Van
Doren. Copyright © 1940, McGraw-Hill Book
Company, Inc. Used with permission of
McGraw-Hill Book Company.*

rings on its sidewall.[20] Oil companies ex-
ploited the fad. According to an advertisement for Sunoco, "A falling drop of water led
the way to STREAMLINED DESIGN and every
drop of BLUE SUNOCO leads to STREAMLINED
ACTION." Motorists learned "whether your
car is streamlined or not, Blue Sunoco will
give you streamlined speed and economy—
because its refining process eliminates those
undesirable parts of petroleum which retard
swift acceleration, high speed and knockless
power." Whatever the gasoline's merit, the
ad was appropriate—appearing three days
after the *Zephyr* glided into the Century of
Progress Exposition.[21] Some years later a
minor designer even "ventured to suggest"
to shoe manufacturers "that automobile
streamlined fronts could be adapted to some
shoe designs."[22]

Consumer engineer Egmont Arens, quick
to recognize a trend, campaigned for business and government use of streamlining as a
recovery slogan. Although his efforts produced little, they did reflect public response
to the concept and foreshadowed streamlining's use as a coherent design style. In

January 1934 he offered to promote the Airflow with a pictorial magazine called SPEED,
which "would carry news in pictures of this
modern world playing up wonders of engineering and science and dramatizing the stir-
ring rebuilding of civilization we are now
witnessing." SPEED would demonstrate that
the company was "well in the vanguard of
this new world" and educate people "to
accept radical departures from accepted design" by demonstrating that "all fast moving
things on earth eventually adopt the teardrop."[23] When Chrysler declined his offer,
Arens developed a slide lecture—"See
America Streamlined"—for delivery to civic
groups. Returning from a swing through the
Midwest in November 1934, he described "a
welling up of national enthusiasm and
energy" behind streamlining.[24] To Arens it
became "THE WORD of the great liberation"
because it seemed "to release the wishes and
hopes of people in all walks of life whose
will and whose energy have been chained
down by the circumstances of the depression." Streamlining had become "a natural
phenomena [*sic*] . . . a crystalization [*sic*] of
mass psychology that cannot be laughed
off."[25]

After addressing the citizens of Owatonna,
Minnesota, Arens fired off an exuberant telegram to President Roosevelt: "Arriving Washington Sunday with lantern slide talk called
Streamlining for Recovery showing Streamlined Aeroplanes, Zeppelins, Ocean Liners,
Trains, as well as Streamlined Trees, Flowers,
Whales, Diving Girls, Refrigerators, Houses,
Gadgets, Womens Fashions Stop Urging Industrialists to Streamline their business and
plant for Recovery Stop Delivered in Min-
neapolis and Chicago this talk has aroused
such eager response suggest it might be used
for Recovery slogan by all Alphabetical Divisions Stop Word Streamlining has captured
American imagination to mean modern, efficient, well-organized, sweet clean and beautiful Stop You would love it Stop. . . ."[26]

Arens's telegram was referred to the Federal Housing Administration, which arranged

for him to meet with businessmen on an Industries Sales Committee. They hired him to design a truck caravan to tour the country promoting kitchen appliances and home furnishings. His report on the Better Homes Caravan focused on streamlining. The trucks, Arens argued, "should have somewhat the same aspect of one of the streamlined trains." The house interiors displayed in the trucks "should be 'Streamlined for Selling'—eye resistance eliminated . . . making it always easy for folks to sign the order pad." [27] Streamlining, he admitted, was "a rather intangible thing," but "used as a selling tool," it would "have very remarkable results." By steering the caravan along the current of enthusiasm he had noticed, the committee would transform it "from a mere commercial enterprise into a movement of some national significance." The caravan would "capture people's imaginations" and contribute to national morale by carrying "some educational and inspirational freight in addition to the poundage of bath tubs, refrigerators, washing machines and what have you." [28] With the arrival of streamlining on the American scene, designers could create a style that would express "this peculiar genius of the American people to be going places—and be going there fast." [29]

Raymond Loewy echoed Arens's enthusiasm when he told a reporter that streamlining represented "a state of mind"—"the perfect interpretation of the modern beat." But in the smooth, flowing surfaces of the style he discerned an appeal that contradicted desire for rapid change. Streamlining, Loewy asserted, "symbolizes simplicity—eliminates cluttering detail and answers a sub-conscious yearning for the polished, orderly essential." [30] The style significantly gained popular acceptance when applied to static objects and buildings as well as to trains and autos. The American people, shocked by a depression that was in part caused by technological change, did not desire limitless progress alone. They also desired an orderly frictionless society whose scientists and engineers would eliminate uncertainties and complexities—a society that, once established, would provide security by not changing. The repeated assertions that streamlining's teardrop represented the ultimate, most functional form for a vehicle contained a logical corollary—that after the streamlined style transformed the environment, society would be housed in a static, eternally perfect machine. Society would move through time with as little surface disturbance as a teardrop auto moving through air.

Streamlined motifs appeared with increasing frequency in all types of design. The automobile industry was not far behind the railroads. Experimental cars like the Dymaxion and radical production cars like the Airflow stimulated publicity, but automakers capitalized on the fad with minor body changes. Automobiles of the early thirties, introduced before streamlining became a sales concern, had some rounding of corners, but generally they had rectangular lines. From 1935 to 1937 the impact of streamlining on auto body design became apparent. In the typical production auto the roof line became more rounded, flowing continuously from hood through front window posts to curved rear window posts and a somewhat tapered rear. The roof became an umbrella under which the windows became smaller umbrellas, thanks to center posts that flared out as they extended up. Fenders became more solid-looking, particularly behind the wheels, as they took on definite teardrop outlines prior to being incorporated within the body. The most radical changes occurred in front-end treatment. Headlights were faired into front fenders. Hood and radiator became a single integrated unit as car design tended more toward rounding and enclosing. To customers with no knowledge of aerodynamics, such cars appeared streamlined when compared with the angular boxes of a few years earlier.

Other types of vehicles took on streamlined forms. As early as May 1934 the *New*

112. Truck for Labatt Ale of Canada, 1937. Food
Engineering.

*113. Loewy. Greyhound bus, c. 1939. Raymond
Loewy.*

York Times ran a photograph of an oil truck
designed by a Fruehauf engineer. Its cab
had straight lines, but the trailer, rounded on
all sides, looked like a floating whale as its
top half bulged up and then subsided to the
rear.[31] By 1937 such truck forms became
commonplace, particularly in consumer-
oriented businesses that were willing to
trade ease of access to a truck's storage space
for the advertising value of a streamlined
vehicle (see figure 112). Bus companies
could not afford to be so generous with
space. Around 1939 Loewy began a long
association with Greyhound by designing an
essentially rectangular bus (figure 113). It
had obligatory rounded corners, but most of
its streamlining derived from its color
scheme, which included a stylized teardrop
painted around each wheel opening.

Most obviously streamlined were the airplanes of the thirties. In 1933 Douglas developed a single DC-1, a prototype of the DC-3,
which went into production in 1935 and
began commercial service in 1936. As a historian noted recently, these planes and the
Boeing 247, introduced in 1933, marked the
appearance of the completely streamlined
airplane with their "hemispherical noses,
tapering cylindrical fuselages, and swept-
back wings and stabilizers." [32] By the late
thirties industrial designers were consciously borrowing airplane forms for designing
static objects, particularly small household
appliances. Robert Heller, a young designer,
admitted that inspiration for his Airstream
fan (figure 114) for the A. C. Gilbert Company came from the airplane propeller.[33]
Unlike most previous fans, which had four
blades, the Airstream had two, and its
smooth ellipsoid motor housing resembled
an airplane motor shell. A mixer designed by
Heller duplicated the same forms and replaced a model whose cluttered parts were
visually unintegrated. Loewy's streamlined
pencil sharpener (figure 115), the epitome
of airplane-derived design, had a teardrop
housing of polished chrome that also re-

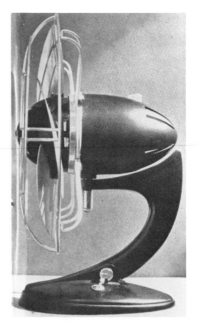

*114. Robert Heller. Airstream fan for A. C. Gilbert
Co., c. 1938.* Art and Industry.

*115. Loewy. Pencil sharpener, 1934. Raymond
Loewy.*

called an airplane motor, especially when its handle was in motion.

Streamlined products also reflected borrowings from vehicles other than the airplane. The "automotive effect" of the V-shaped door on Loewy's 1937 Coldspot refrigerator has been noted. The Hoover 150, an upright vacuum cleaner designed by Dreyfuss in 1936, took its lines from the streamliners (see figures 116 and 117). Made of plastic and lightweight alloys and colored "stratosphere grey and blue," it had a broad, flat projecting base that looked like a stylized cowcatcher, surmounted by a looming teardrop motor housing resembling a locomotive cab—an effect heightened by a "headlight" for dark closet corners.[34] Such direct borrowings had to be limited: appliances whose parts could not fit in a single housing required more complex abstract forms. The Hobart Streamliner (figure 118), a meat-slicer designed by Arens for delicatessen use, exhibited a plethora of gleaming rounded forms. They were not well integrated, but there was hardly an angle to be found.

By the end of the decade, curvilinear forms had nearly replaced modernistic angular forms in product design. In its 1935–1936 catalogue Sears offered a few streamlined items, including a bathroom scale with rounded corners and a teardrop dial housing. Trumpeting it as "America's Most Beautiful Scale," Sears proudly announced that its "sleek, streamline headpiece is *chrome plated*."[35] Three or four years later such puffery would be out of place, as products like toasters, mixers, cameras, and radios were given rounded lines as a matter of course. Van Doren's plastic Air-King radio case (figure 60) had exemplified the stepped, setback, vertically oriented forms common in the early thirties, but by the end of the decade most plastic cases, though generally rectangular, were horizontally oriented and had rounded edges and corners. Even wooden radios exhibited a wraparound flow of continuous graining up one

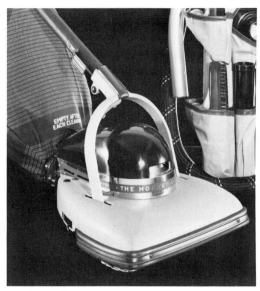

116. Dreyfuss. Hoover 150 vacuum cleaner, 1936. Courtesy of the Hoover Company.

side, along the top, and down the other side—made possible by application of veneer to molded composition materials. Products that retained the angular step motif, such as a home thermostat designed in 1937 by Ted Hess for the Perfex Controls Company (figure 119), did so with softened edges and curvilinear elements.

The triumph of streamlining as a product style appeared complete in 1940 when Van Doren issued his manual *Industrial Design*. Aware of the criticism of streamlining static objects, he pointed out to young designers that "the manufacturer who wants his laundry tubs, his typewriters, or his furnaces streamlined is in reality asking you to modernize them, to find the means for substituting curvilinear forms for rectilinear forms." To be successful, products had to "conform to the current taste, or fad if you will, for cylinders and spheres or the soft flowing curves of the modern automobile in place of the harsh angles and ungainly shapes of a decade ago." If "borrowed streamlining" appealed to "popular fancy," he concluded, then use it.[36] From another source, equally an object of popular fancy, came confirma-

HIGH SPEED COMMUTING CAR – Ideal dimensio

tric or electric. Des

*117. Loewy. Speculative rendering of high-speed
commuting car, 1935.* Creative Design.

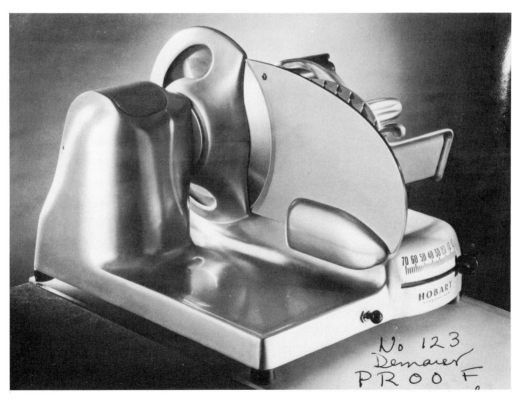

*118. Egmont Arens. Hobart Streamliner meat
slicer, c. 1935.* Arens Collection.

119. Ted Hess. Thermostat for Perfex Controls Co., 1937. Reprinted by permission of Modern Plastics Magazine, McGraw-Hill, Inc.

tion of streamlining's success. As Frederick Lewis Allen noted in his review of the thirties, the previous decade's ideal woman, the flapper, had a figure running "straight up and down—no breasts, no waist, no hips"—except perhaps for a few protruding angles. But in the thirties "Mae West's curves became a national influence."[37]

Associating streamlining with a seductive actress is not entirely facetious. In 1938 John Vassos attacked Hollywood for providing "every demi-monde and every kept woman" in films "with a modern interior," while "all the virtuous girls were surrounded with Colonial homes."[38] Films merely exaggerated reality. Contemporary observers noted that Americans installed objects of modern design in kitchens and bathrooms while retaining imitation period furniture in living rooms and bedrooms.[39] In the average kitchen, however, a streamlined toaster or mixer remained isolated in a clutter of older,

eclectically styled objects. Only the wealthy could afford to replace at once all the furnishings of a kitchen, not to mention those throughout a house. Only the wealthy could aspire to a coherent machine-age environment, starting perhaps with a house by Richard Neutra or partners George Howe and William Lescaze. For the average person the streamlining of an appliance functioned primarily at the point of purchase, where his or her attention was fixed without distraction on a contemplated acquisition. A housing suggestive of an airplane or a streamliner reminded a customer that this particular gadget provided, like those more exalted machines, the benefits of modern technology. In addition, the familiar surroundings in which the new appliance was placed domesticated it and made its users feel at home with technological change.

Despite their domestic conservatism, Americans accepted machine-age styles in public architecture. The same observers who lamented retention of period furniture in homes applauded the public's enthusiasm for modern shops, restaurants, and theaters. Their appeal resembled that of the movies, which provided a brief escape from Depression cares. It was not surprising that on many small-town streets modernized in the thirties, the most impressive facade belonged to a cinema, its marquee an aluminum-banded parabolic curve (figure 120). Public architecture with modernistic or streamlined motifs provided sets in which people could act as if the technological future that they symbolized had arrived. Surrounded by the new architecture's polished impersonality, even a poor citizen could escape momentarily into the future. The very impersonality of the streamlined style limited its architectural use primarily to public and commercial architecture. Most people surrendered willingly to projections of a mass society as smoothly functioning as a frictionless machine, but they sought to maintain individuality in their homes.[40]

Several years before the new trains made

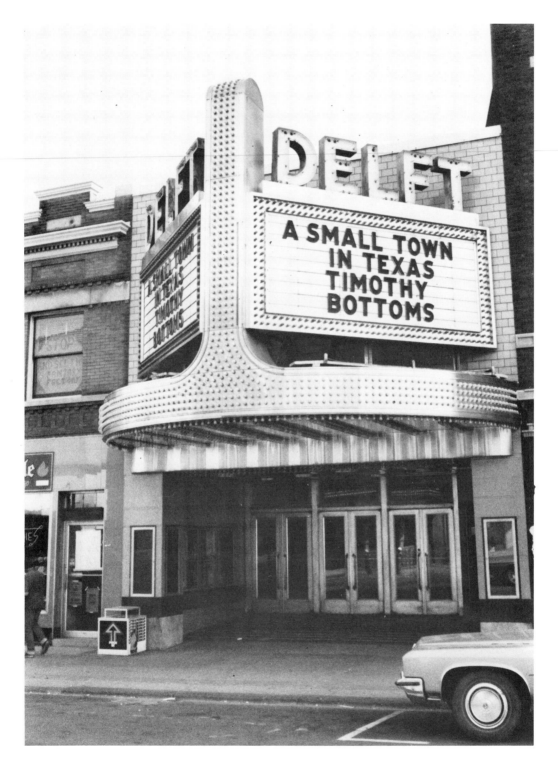

120. *Delft theater, Marquette, Michigan. Author.*

121. Hut's diner, Austin, Texas. Author.

streamlining a fad, curvilinear elements had softened modernistic architecture's angular lines. Derived from Europeans like Eric Mendelsohn and J. J. P. Oud, streamlined architectural forms coalesced with "aerodynamic" forms for vehicles and products to make a coherent style. In 1934 an aeronautical engineer at the University of Michigan suggested using aerodynamic forms in skyscrapers to reduce cost by eliminating structural elements previously needed "to withstand wind pressures."[41] Such considerations had no impact outside the cities of Flash Gordon. Nor were streamlined buildings as representational as streamlined products. Except for an occasional diner (figure 121) or gas station surmounted by an apparent airplane tail, most buildings remained abstract, their curved forms and

metal trim merely suggesting vehicular design.

Mendelsohn admitted the American automobile's impact on his work, but his buildings expressed that machine's spirit as he perceived it rather than directly representing its forms. His experience foreshadowed the fate of American streamlined architecture. Only two of his structures, the Einstein Tower and the Universum theater, captured the dynamic plasticity of sketches from his early notebooks. His commercial facades of the twenties paled by comparison. In America to a greater extent—excepting Frank Lloyd Wright's Johnson Wax administration complex (1936–1937), a few buildings at the New York World's Fair of 1939, and Eero Saarinen's work of the late fifties—streamlined architecture was limited to surface

122. Raymond Hood. McGraw-Hill building, New York, 1931. Cervin Robinson.

treatment of ordinary facades and interiors.[42] By rounding corners inside and out, by emphasizing exterior flow with horizontal bands of glass or metal, and by directing interior flow with curved railings, pinstriped Flexwood paneling, and troughs of indirect lighting, an architect could imply that the complex processes of a bank, a government agency, a factory, a department store, or even a bus terminal, were proceeding smoothly and efficiently, despite evidence that this was not always the case during the Depression. Streamlined architecture, once again, provided a wish-fulfilling set for a secure, effortless future whose complexity would run no deeper than a polished surface.

Two skyscrapers of the early thirties with Mendelsohn-inspired streamlining on lower floors marked the style's appearance in the United States. Raymond Hood, former popularizer of modernistic architecture, provided New York's McGraw-Hill building, completed in 1931, with an entrance (figure 122) whose sides curved in to a bank of doors. The curved sides, a light metallic blue cut by bands of dark turquoise outlined in light blue tubing and split by flat strips of brass, provided a flowing motion not so obvious in the horizontal but right-angled bands of floors above. More impressive was the Philadelphia Saving Fund Society building (figure 123), designed by Howe and Lescaze and completed in 1932. William H. Jordy has skillfully analyzed functional elements of its setback, T-shaped office tower, but at street level expressionist packaging prevailed.[43] The three lower floors formed a single horizontal curve wrapped around two sides of

123. George Howe and William Lescaze.
Philadelphia Saving Fund Society building,
1932. London Studio.

the building. Above display windows of shops on the first floor the architects placed a sheer overhanging wall of reflective charcoal granite, broken by a vast band of glass, flush with the surface, indicating the second floor bank lobby. No projection or detail disturbed the building's smooth horizontal flow as experienced from the street.

As the decade progressed, streamlined forms appeared throughout America, in small towns as well as cities. Inward curving entrances became common, glass for store fronts and brick or stucco for offices, apartment houses, and bars. Often a shallow streamlined overhang of metal or concrete extended over such entrances, complementing the flowing lines below. During modern-

ization department stores sprouted wide aluminum-banded streamlined marquees. Proprietors of independent service stations attempted to attract customers by tacking streamlined canopies onto their buildings, possibly in imitation of Teague's Texaco design but often hopelessly ill-proportioned. Bars, nightclubs, and restaurants, whose clientele desired privacy, had solid stucco walls and rounded corners accented with horizontal bands of translucent glass bricks. Such entertainment meccas (see figures 124 and 125) often displayed a painted stucco marquee, half a foot thick and three or four feet wide, encircling the building's exposed sides, its curve echoing the continuous wall. Double doors of bright metal, each inscribed

124. *Hill's restaurant, Galveston, Texas, 1941.*
Author.

125. *Nightclub, Rosenberg, Texas. Author.*

with a semicircular window, typically completed such a design. Smooth, gleaming, and rounded, such streamlined buildings symbolized the future, but a discerning observer would have seen it was all for show. In the back alley, where appearances made no difference, the curving metal and concrete yielded to squared-off functional construction. The streamlining of society through its architecture remained as cosmetic as the streamlining of its autos.

Industrial and government architecture also showed a penchant for streamlining surfaces of public areas. Albert Kahn, a prolific industrial architect, gave superficial attention to the new style in designing administration areas of his plants. Although he adhered to rectilinear forms throughout production areas of his factories, he provided main entrances with inward-flowing walls, bands of glass brick, and curved overhangs. Inside, a visitor might find a rounded plywood reception desk and curvilinear walls accented above by luminous bands of channeled light.[44]

A remodeled courthouse in Nevada City, California, provides an example of selective streamlining of public surfaces, both exterior and interior. Built in 1864 of granite and brick to resemble an Italian Renaissance palace, it was reworked in 1938. Two low wings surfaced with smooth concrete curved back symmetrically from a modernistic central section. All offices adjoining the front walls of these wings enjoyed curved interior spaces. But in addition the main courtroom, located well within the structure and surrounded by rectilinear halls and rooms, possessed rounded corners, probably of Flexwood. Even the process of justice yielded to the efficient flow of modern times.[45]

Similar examples of streamlined motifs appeared throughout public architecture of the thirties. WPA buildings relied on a rugged adaptation of the modernistic style, whose verticality proved appropriate to the physical and psychological elevation of public buildings. Most of them, however, had flowing curves around entrances, in important interior spaces, and in stair railings and drinking-fountain wells. Even TVA powerhouses, usually monolithic rectilinear structures of undressed concrete, offered streamlined curves in lobbies and reception rooms. The walls of a lobby in the Pickwick Landing powerhouse (figure 126) made few right angles but instead flowed sharply around corners, into recessed doors of polished aluminum, and into a stairwell. Tubular aluminum furniture complemented the flow of the walls. Hardly an impressive room, its design probably issued from the drawing board of a young architect who employed available motifs. By the end of the decade the streamlined style had become part of the architectural vernacular.

Curved lines in architecture were intended to aid traffic control. Inward-flowing lines at a store's entrance led a pedestrian's glance to its interior and increased the chance that he or she would enter. Similar devices in public buildings drew attention to the door of an important office or meeting room, a bank of elevators, or a stairwell. And rounded corners supposedly promoted efficient movement of people through corridors. Streamlined forms also predisposed people to feel that public functionaries were moving quickly and efficiently to process their business.

In the final analysis, streamlined forms served a more significant psychological function. A rounded exterior facade attracted attention to a building, but it also provided a sense of enclosure. However incomplete a curve, it invites completion. And inside a building, curving walls, often banded with polished metal, complementary bands of indirect lighting, luminous walls of glass bricks, their illumination so altered that it might have come from hidden electric bulbs rather than the sun, and the faint hum of ventilating equipment, all conspired to produce an impression of enclosing one in a totally controlled artificial environment.

126. *Lobby of powerhouse, Pickwick Landing,*
Mississippi. Pencil Points.

Future sets created by architects in the thirties reduced uncertainty nearly to zero. Although the association of streamlining with speed suggested endless progress into the future, its smooth forms also paradoxically suggested maintenance of static perfection.

Nothing expressed this idea better than a streamliner designed by Henry Dreyfuss in 1938 for the New York Central. Even its name, *20th Century Limited*, inherited from earlier trains, ironically implied a constricted future. Dreyfuss controlled every detail of appearance, from a smoothly shrouded gray steam locomotive (figure 127) and gray cars with blue and silver stripes to its interior finish, tableware, and match covers. Inside, metallic shades of blue, gray, and rust-brown predominated. Common thirties' materials—polished alloys, Flexwood, cork paneling, and Formica—provided a crisp modern finish. Each public car contained partitions to break up the tunnel effect common to railway coaches. And each was rounded, not at the roof line as in tubular construction, but

at each end. The observation car (figure 128) most coherently embodied streamlined design. Between each pair of long windows ran a rounded gun-metal column from floor to ceiling. Along the wall between every two columns was a blue leather couch divided into four seats by wide armrests with bright metal striping. Gray carpeting, gray leather walls, circular metal ashtrays, and a recessed curve of light reflecting onto the edges of the ceiling completed the car's machined look. At each end Dreyfuss placed an inward-curving couch and circular table surmounted by a large photomural, one showing the skyline of Chicago, the other New York, the terminal points of the train's run. Behind the New York mural was the glass-enclosed tail of the car, replacing the traditional open-air observation platform and equipped with two outward-facing semicircular couches from which passengers could see where they had been.

An observer watching the train sweep by might think it presaged the future. But those who chose to "see America streamlined"

127. Dreyfuss. 20th Century Limited *locomotive,*
1938. Dreyfuss Archive.

128. Dreyfuss. 20th Century Limited *observation car.* Architectural Forum.

by purchasing a ticket on the *20th Century Limited* were already experiencing the future. Their artificial environment, compact and comfortable, resembled the future they had learned to imagine. Fully enclosed by glass and rounded surfaces, protected from weather and engine soot by air-conditioning, and lulled by radio music, they could forget they were traveling as they experienced a rubber-cushioned ride softened further by roller bearings and "tightlock couplers." Scenery rushed by outside like a fan-driven wind over a teardrop model suspended motionlessly in a wind tunnel. Within the cars time stopped. Their streamlining encapsulated a moment of the perfect future—only an ideal for a chaotic society but realized on the scale of a train.[46]

Widening Control

Americans accepted streamlining so completely that in common usage the word came to mean simplifying any object or process to increase economy or efficiency. But a significant minority found nothing efficient about a design style that they thought functioned only to stimulate sales by rendering serviceable goods "obsolete" in appearance. Ignoring as crass the consumer engineer's argument that streamlining promoted an efficient industrial system by bringing distribution in line with production, they defined efficiency in design as perfect fit of product to intended use. Any object whose form was intended to attract purchasers by reference to a more glamorous machine reflected bad taste. Most functionalist criticism of industrial design and streamlining emanated from the Museum of Modern Art in New York. Industrial designers, well aware of the uneasy balance between art and industry, must have resented the MOMA attack. They had, after all, gone into business blessed by Dana and Bach, museum advocates of high aesthetic standards in mass production.

In April 1934 the Museum of Modern Art presented its version of industrial design in "Machine Art," an exhibit organized by Philip Johnson. As outlined in the catalogue, which began with an epigraph from Plato stating that absolute beauty of form derived from "straight lines and circles, and shapes, plane or solid, made from them by lathe, ruler and square," the museum's position on machine design differed little from Teague's.[47] Alfred H. Barr, Jr., claimed in the foreword that "the role of the artist in machine art is to choose, from a variety of possible forms each of which may be functionally adequate, that one form which is aesthetically most satisfactory." The true artist, he continued, "does not embellish or elaborate, but refines, simplifies and perfects." Johnson clarified the museum's position on contemporary design in his short essay, "History of Machine Art." As opposed to the "irregularity, picturesqueness, decorative value and uniqueness" of handicraft, machine art exhibited "precision, simplicity, smoothness, reproducibility." But legitimate American designers were no longer overcoming the handicraft tradition. They were fighting two misguided approaches to ma-

chine art: first, the "neo-classic trappings and bizarre ornament" of "a 'modernistic' French machine-age aesthetic"; second, "styling" and ideas like streamlining that "often receive homage out of all proportion to their applicability."

Johnson's selections revealed his position clearly. Those few styled products shown—such as Kodak cameras and Taylor barometers—were dismissed by functionalist critic Catherine Bauer as "among the more painful exhibits."[48] But another critic observed that most objects revealed "the unconscious result of the efficiency compelled by mass production" rather than "the directed art of an industrial designer."[49] Johnson emphasized purely industrial objects like sections of wire cable, springs, ball bearings, insulators, gears, and pistons—displayed as unique works of sculpture. Through this exhibit MOMA signaled to industrial designers that it sanctioned only work artfully derived from pure engineering considerations. Five years later, when the museum celebrated its tenth anniversary with a retrospective exhibit, its Industrial Design section pointedly contained only Fuller's prefabricated bathroom and chairs by Le Corbusier, Mies van der Rohe, Marcel Breuer, and Alvar Aalto.[50]

Not all design purists disliked streamlining. For example, Lewis Mumford rhapsodized about vehicular streamlining in his *Technics and Civilization*. The airplane, the first new machine of the efficient neotechnic phase of civilization, had required development of nonmechanical, organic forms derived from the fish, just as the automobile eventually would emulate the turtle. Some machines became less mechanical in form as engineering and aesthetics, at odds during the chaotic paleotechnic phase, now coalesced. This new organicism seemed natural to Mumford because both the sense of beauty and the recognition of efficiency derived "from our reactions to the world of life, where correct adaptations of form have so frequently survived."[51]

However, such considerations hardly applied to streamlining objects that, unlike turtles and fish, did not move. Whatever its expressionist significance or its cash value, streamlining of static objects was not required for efficiency. Writing to Geddes in 1934, director Barr of the Museum of Modern Art referred to streamlining as "an absurdity in much contemporary design." Singling out for abuse the "streamline pencil sharpener by one of the highest paid industrial designers" (figure 115), Barr noted that such "blind concern with fashion" made it "difficult to take the ordinary industrial designers seriously." Thus, he and his museum associates "played up the anonymously designed industrial object rather more than the object which shows evidence of 'styling.'"[52] Another museum official, John McAndrew, who inherited the department of architecture and industrial art from Johnson late in the thirties, typically resorted to sarcasm when discussing streamlining by observing that "streamlined paper cups, if dropped, would fall with less wind-resistance," but they were "no better than the old ones for the purpose for which they are actually intended, namely drinking."[53] Similar witticisms for years characterized the MOMA attack on streamlining and the industrial designers who used it.

The designers, who sought aesthetic respectability as well as financial remuneration, justified static streamlining. Whatever its aerodynamic value, streamlining marked the emergence of a national American style that derived from neither French modernistic design nor German functionalism. As early as 1928 Henry-Russell Hitchcock, Jr., predicted that modernism would exceed its status as an imported "fashion" to become "a vital and developing style" shaped by Americans.[54] That same year the Art Alliance of America announced its intention of proving "that when American industry and native designers cooperate the quality of the product is immediately enhanced."[55] Four years later, when the Rockefeller-funded National Alliance of Art and Industry held its first industrial design exhibit at the Art Center,

the organization stated as its purpose "the forming of a national taste and a national school of design."[56] In 1934, when its annual show moved into the RCA building, a preliminary announcement aimed at designers and manufacturers headlined a similar purpose: "TO CELEBRATE THE EMERGENCE OF AN AMERICAN STYLE."[57] At some point the exhibit's planners, who included all the leading industrial designers, contemplated using streamlining as an organizational concept. This idea was dropped, but streamlining remained a central focus. Not only did the exhibit feature Loewy's pencil sharpener, but one critic even identified streamlining as the new national style. Americans might be "lagging somewhat behind Europe," but they were "developing a national style" marked by disappearance of "sharp edges and corners." This critic observed "that numerous curved forms are taking their place in the commercial designs of utilitarian products."[58] National pride thus dictated acceptance of the streamlined style as a declaration of independence from European designers.

Purist critics concluded from such assertions that American industrial design continued to lag behind. A different rationale for streamlining static objects seemed more likely to satisfy those who demanded functional design. László Moholy-Nagy, a Hungarian who taught for five years at the Bauhaus before coming to the United States in 1937, defended streamlined design as economical because "casts, molds, stampings as well as finishes . . . could be more easily produced" by eliminating square corners and edges.[59] American designers already knew streamlining's technical advantages. When bending sheet metal for a heater or stove, one had to leave a curve to avoid weakening the metal.[60] The plastics industry also benefited from rounded forms. A mold with sharp corners had to be hand-polished, but a rounded mold could be polished by machine. Molding compounds flowed more evenly in a curved mold. Finally, the finished product, perhaps a radio

129. Plastic radio cases, c. 1937. Reprinted by permission of Modern Plastics Magazine, McGraw-Hill, Inc.

cabinet (figure 129), would be stronger and would have no corners to be chipped in transit.[61]

Streamlining's most sincere argument had nothing to do with manufacturing processes. Even a bathroom scale's teardrop dial housing represented valid design because it shared by association in the mystique of the streamlined transportation machine. As a common style, streamlining contributed to a coherent environment. As the Cheneys observed, "each article or object . . . falls into harmonious relationship with all else, becomes integral to the intangible, pervasive twentieth-century atmosphere." This common style provided evidence of "a new subjective force active in the machine-age world, imposing itself upon the external environment as a spirit of order." The Cheneys accepted "the streamline as [a] valid symbol for the contemporary life flow, and as a

badge of design integrity in even smaller mechanisms, when it emerges as form expressiveness." As streamlined designs became prevalent, approaching universality, the "machine-conscious mind" would begin "to relate all such products of scientist-artist design back to the most conspicuous symbol and inspiration of the age"—the airplane—"as the reverent medieval mind related everything to the symbol of the cross."[62]

Teague, too, saw a "new order emerging" as designers carried out a "tremendous job of reconstruction" necessary after the "war" of the industrial revolution. Permeating this new design order Teague found the "constant ratios of proportion" and "the quality of line which we find most highly developed . . . in a Douglas transport plane, where you see the same type of form repeated in the engine and in the fuselage, in the wings and the tail—the same line recurring again and again; that long line with a sharp parabolic curve at the end, which we have come into the habit of calling 'streamline.'" To Teague streamlining static objects seemed appropriate not only because repeating this parabolic curve produced harmonious relationships among objects great and small. In addition, the parabolic curve with "a long backward sweep" seemed "characteristic of our age" because, he concluded, "we are a primitive age, a dynamic people, and we respond only to the expressions of tensions, of vigor, of energy."[63] Teague thus based the validity of streamlining for all objects, whether they moved or not, on its ability to express the spirit of the age. Even Nikolaus Pevsner, the first historian of functionalism, looking back from 1960, admitted that the "bogus streamlining" of the thirties was "functionally unjustified but emotionally justifiable."[64]

The American public did find emotional satisfaction in streamlining. Otherwise, the style would not have succeeded in the marketplace. But pinning down reasons for its success is difficult, requiring intuitive feel for the style and a sifting of conflicting rationales introduced by designers and critics to justify or explain it. Observers did not even agree on the significance of the parabola, which Teague found at the core of streamlining. In 1928 an architecture instructor at the University of Michigan, Francis S. Onderdonk, Jr., praised the parabola as a dynamic form in terms similar to those used later by Teague. Onderdonk's eccentric volume, *The Ferro-Concrete Style*, included photographs of buildings ranging from the Mission style to the International Style, but he reserved greatest praise for Eugène Freyssinet's monumental parabolic dirigible hangar completed at Orly Field in 1916 (figure 130). Onderdonk found this use of the parabola structurally efficient. But he also insisted that "psychological reasons, based on the extension of our limits in time and space which create a new attitude towards the infinite, seem to demand the use

130. Eugène Freyssinet. Dirigible hangar, Orly Field, Paris, 1916. Sheldon Cheney.

of the semi-infinite curve, the parabola, in our age." Since a parabola is part of an ellipse, the path that "comets follow . . . thru limitless space," use of parabolic forms "gives the Ferro-Concrete Style something of the infinite swirl and swing of the Universe" and embodies continual extension of "boundaries in man's struggle with the infinite."[65]

For architect Claude Bragdon—whose treatise *The Frozen Fountain* exceeded Onderdonk's in eccentricity and mysticism —the parabola did not signify mankind's limitlessly expanding conquest of space and time. Instead, it symbolized the somehow comforting fact that human beings are fated to build and strive, to push upward, only to yield in the end to the pull of gravity by falling back. The human race itself, according to Bragdon, is like water shot up in a fountain. Individuals, like single drops of water, "trace each a shining parabola in space-time and subside again into the reservoir of life." Bragdon conceived of architecture as a "frozen fountain." He described "the white vertical masses of the Waldorf-Astoria" as "a plexus of upward rushing, upward gushing fountains, most powerful and therefore highest at the center, descending by ordered stages to the broad Park Avenue river." Just as life itself, beyond the chance circumstances of individuals, remained locked into an unchanging pattern, architecture provided Bragdon with an image of

131. Loewy. Rendering of K4S locomotive for Pennsylvania Railroad, 1936. Raymond Loewy.

"time frozen into eternity."[66] While Onderdonk found in the parabola a promise of limitless progress into the future, Bragdon considered the form an image of an eternally repeating process and ultimately of stasis. Neither of these mystical architects addressed the problem of the expressive significance of streamlining, but they did state it.

It would seem obvious to regard streamlining as a style of a society whose developing technologies propelled it at an accelerating rate into the future. Like the Burlington *Zephyr* leaving its terminal behind, technological society would soon leave the Depression in the past. In 1934 journalist Harold Ward described aerodynamic progress "towards a new era of speed." By eliminating "irrelevancies, excrescences, trivial ornament, time-wasting and energy-consuming projections and gadgets," designers of streamlined vehicles provided "less weight, more power, less effort, more results"—in short, greater speed. These achievements mirrored developments in other technological fields, in which engineers were "learning 'what to do *without*.'" The word speed had "become a synonym for Progress: some would even regard it as defining that ambiguous conception." Ward saw no end in sight but projected a continually receding "Ultima Thule to whet the imagination."[67]

The image of a streamlined society speeding into the future like a rocket suggests masculine sexual associations. Not surprisingly, recent commentators have interpreted streamlined vehicles as expressions of masculine power. Richard Pommer, in a brief review of Loewy's work, has characterized a rendering of the designer's K4S Pennsylvania steam locomotive of 1936 (figure 131) "as a gigantic decorated phallus." Such imagery, he concludes, was "inherent in streamlining."[68] And Donald J. Bush has written of aerodynamics in general as "the science of penetration."[69] If these interpretations are accurate, then streamlining embodied society's urge to pierce the future and transcend itself continuously; or, in more concrete terms, to accelerate its technological development and augment its material well-being without end.

These interpretations do not seem complete. Loewy's locomotives certainly appeared phallic, but Geddes's teardrop automobile did not. An egg, broad end first, provides a more apt organic analogue of the teardrop, which was recognized in the thirties as the ideal streamlined form. Bush seems to admit this in his misleadingly entitled section, "The Science of Penetration," when he cites Sir D'Arcy Wentworth Thompson's treatise *On Growth and Form*, published in 1917. Thompson theorized that natural forms, including the complex ones of animals, derive from gradual adaptation to forces acting upon them from without. Bush quotes with approval Thompson's assertion that an egg in the oviduct of a hen can be seen as "a stationary body round which waves are flowing," those waves being the muscular contractions that give the egg a shape of least resistance. Bush also quotes Thompson's observation that snowdrifts and sand dunes provide "endless illustrations of stream-lines or eddy curves," evidence that a "stream [of air] tends to mold the bodies it streams over, facilitating its own flow."[70] Geddes probably had Thompson in mind when he included in *Horizons* a photograph of an elongated teardrop ice flow to introduce his discussion of automotive streamlining.[71] Even Geddes, a prophet of speed, recognized the teardrop as a form passively shaped by outside forces. And Douglas Haskell, tracing the evolution of the car "from automobile to road-plane," praised streamlining as "a superior approach to the whole problem of design" because "it coaxes nature, yields, guides, and adapts; it is the opposite of 'conquest' by clumsy attempts of sheer force."[72]

Bush does not go far enough when he discusses Thompson's egg. In addition to expressing an aesthetic of penetration representative of a technological thrust into the future, the streamlined style of the thirties also paradoxically embodied and expressed a desire for stasis engendered by the Depression's social chaos. Technological progress had deepened the Depression by extending the industrial machine's capacity to produce goods beyond society's capacity to consume them. The future ideal society would have machines undreamed of, but there would be no place for technological acceleration with its potential for chaos. Society's functions would be as efficient and repetitious as the streamlined forms of buildings in which they took place. The streamlined society would be as smooth and uncomplicated as an egg— a mechanical egg, with no possibility of unruly life breaking out from within. John and Ruth Vassos had envisioned this society in 1930 with their *Ultimo*, containing illus-

132. *John Vassos. Visualization of underground city of the future, 1930. From ULTIMO by Ruth and John Vassos. Reproduced by permission of E. P. Dutton.*

trations of streamlined architecture with implications far different from those of Hugh Ferriss's thrusting modernistic skyscrapers (figure 132).

Thus, the streamlined style expressed not only a phallic technological thrust into a limitless future. Its dominant image, the rounded, womblike teardrop egg, expressed also a desire for a passive, static society, in which social and economic frictions engendered by technological acceleration would be eliminated. Streamlining was paradoxically a style of retreat and consolidation as well as one of penetration and forward progress. No industrial designer fully intellectualized a vision of future stasis. But this gained complete expression appropriately enough in the pages of a science-fiction pulp magazine, . *Amazing Stories.* A "nonfiction" article described the future city as constructed not "of steel and stone, of brick and mortar, of wood and nails" but "of seamless, cast and rolled plastic materials of brilliant and beautiful colors." Architecturally, it would be "a city of curves and streamlines, of sweep and rounded beauty." Its streamlined facades would house "a mecca of peace, quiet and contentment, and a wonder-house of science and industry and mechanical coordination." Above all else, the city of the future would provide "Man's Utopia at last." [73] That this static environment would be constructed of plastic is not surprising. Often described as "the hallmark of 'modern design,'" [74] plastic made possible "minute accuracy, strength, permanence, and stability." One designer observed that the future "promises to be a true Plastic Age." [75] But he failed to note that "a plastic is anything *but* plastic, once it has hardened and taken shape." [76]

Few industrial designers would have found this vision in their work, but it was there. They united in considering their major role, apart from eliminating frictions of wind resistance and sales resistance, to be eliminating friction between products and users. Dreyfuss stated this goal as his official motto, and Loewy often asserted that he entered the profession out of personal frustration with goods that refused to work properly. As an observer of the scene noted, industrial designers served "by eliminating the unnecessary, clumsy and poorly designed objects which had cluttered up the lives of the people." [77] Eliminating friction between products and users might be phrased as eliminating uncertainty. Designers attempted to extend a user's certainty of control over his or her environment. Not only could a streamlined housing prevent hair or clothing from being caught in a machine, but its smooth, simple surface could also provide an operator with a feeling of confidence in the face of complexity. "Just as the mind is bored by technicalities it does not understand," a design publicist observed, "so the eye is tired by complex shapes it cannot grasp without effort." [78] The Cheneys argued that streamlining even the smallest and least significant products offered "an

added element of machine-age freedom by widening control over the minutiae of the daily routine." [79] And Lewis Mumford declared that "simplification of the externals of the mechanical world is almost a prerequisite for dealing with its internal complications." [80] Most functionalists would have admitted that both Loewy's pencil sharpener and Howe and Lescaze's Philadelphia Saving Fund Society building contributed to that goal.

When applied to individual products, both industrial design and streamlining increased one's freedom through greater control of minute segments of the environment. But designers often hoped for more. Teague advocated rational planning of society as a whole to reduce friction among its parts. Sheldon and Arens suggested something of the sort with consumer engineering. By manipulating the appearance of products, they hoped to shape the flow of goods from producers to consumers and thus streamline society's production-distribution system. Consumers might enjoy freedom of choice, but their actions would be controlled subtly by the nature of the available goods. The cross-purposes of competing manufacturers using consumer engineering would be reduced if similar techniques of control were applied to society as a whole. The central metaphor of industrial design—eliminating friction or resistance—would then include smoothing away, through social engineering, all potential disturbances, whether of action or expression. The curved entrances and rounded interior corners of streamlined architecture provided intimations of how such engineering might be conducted at the crudest level. The obstinate refusal of most Americans to accept modern architecture for their own houses revealed an intuitive awareness that its smooth impersonality might eventually rob them of individuality. Perhaps they realized, as did British design critic Herbert Read, that "the new [machine-age] aesthetic finds an echo in the new politics, the politics of collectivism." [81]

Industrial designers were not totalitarians and fascists. But the vision of a frictionless society embodied in their streamlined style reflected a positive response to social tensions of the Depression similar to those that produced the destructive concept of a thousand-year *Reich* in Germany. The rapid decline of architectural streamlining after the New York World's Fair of 1939—an event that had seemed to herald streamlining's emergence as the national style—resulted from a general realization of its totalitarian implications. The future social order expressed in streamlining was one of stasis as well as progress. By definition, a frictionless society would be one with no irritations to break its equilibrium. When designers meditated on society's evolution, they tended to think in terms of permanence, just as they conceived of the automobile evolv-

ing toward the ultimate form of the teardrop (figure 133). Despite his description of the current age as dynamic, Teague echoed Mumford, who portrayed civilization's neotechnic phase as an extended period of consolidation. Teague looked forward to a new golden age of peace and harmony in marked contrast to the industrial revolution's severe dislocations, which had culminated in the Depression. He based his philosophy of design on the premise that for a given machine there existed a Platonic form, which it was the designer's task to find and reveal. Extension of this philosophy to society as a whole implied discovery of an ideal form for the ultimate society.

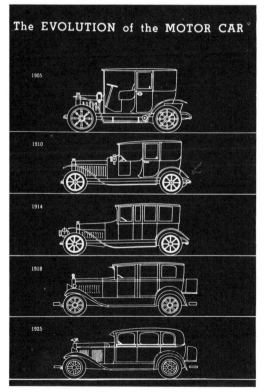

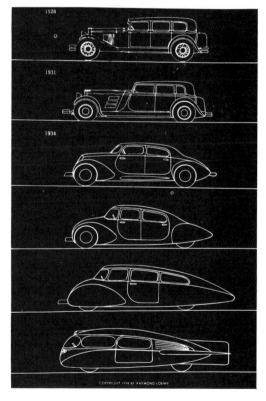

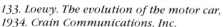

133. Loewy. The evolution of the motor car, 1934. Crain Communications, Inc.

*134. Trylon and Perisphere behind General
Motors building, New York World's Fair of 1939.
Geddes Collection.*

A Microcosm of the Machine-Age World

... the danger of being fantastic in a Wellsian sense, and of being too definitely prophetic, must be avoided.
—*Walter Dorwin Teague*

It will cease to look like a collection of things for sale and reveal its real nature as a gathering of live objects, each of which is going to do something to [us], possibly something quite startling, before [we are] very much older.
—*H. G. Wells*

War in Europe seemed certain when the New York World's Fair opened in the spring of 1939, dedicated to "building the world of tomorrow." Exhibitors and fair-goers alike sought momentary escape in considering, as they had planned, the world of the future. The Century of Progress Exposition had marked the culmination of the modernistic style of New York architects; now the fair of 1939 embodied in its streamlined buildings and exhibits the methods and visions of industrial designers. Teague, who served with six architects on the Board of Design, turned the fair's commercial promoters from an original intention to celebrate the 150th anniversary of Washington's inauguration as president at New York to a goal of envisioning the future society, which he had been projecting through the decade.[1] He and other designers provided "focal exhibits" within major buildings devoted to such subjects as transportation, communication, and science; conceived exhibits for private firms within those buildings; and designed complete buildings, inside and out, for major industrial concerns. Virtually all observers agreed that the fair's greatest hit was the Futurama, designed by Geddes for General Motors, a huge diorama depicting projected superhighways and cities of 1960. It was the industrial designers' fair.

The fair would not have become a reality without help from the one man in America who planned and built on the total scale that industrial designers aspired to but attained only in transitory dioramas. For fifteen years, as head of the Long Island State Park Commission, the New York City Park Department, and the financially and politically independent Triborough Bridge Authority, Robert Moses had put his stamp on the nation's largest city by projecting, planning, *and* building parks, parkways, bridges, and expressways. For more than a decade he had dreamed of creating a park larger than Central Park, in northern Queens just south of Flushing Bay, an area of swampland whose only elevation was an "expanse of stinking refuse known as the Corona Dump."[2] The project became essential to Moses after construction of his Grand Central Parkway, which skirted the area's western edge, in order that motorists need no longer drive through mountains of garbage. When he learned of plans for the fair, he refused to let promoters use any city parks but offered to put his considerable influence behind reclamation of Flushing Meadows as a fair site, provided that it be planned as a permanent city park.

Armed with $59 million in state and federal funds obtained by Moses, the city purchased the site and the fair corporation set to work reclaiming it. Working twenty-four hours a day for nine months beginning in July 1936, teams of laborers filled in the swamp with six million cubic yards of ash and garbage and shaped two lakes, from which a million cubic yards of sludge were taken to become topsoil after chemical treatment. Two new sewage-treatment plants rid Flushing Bay of pollution, while a tide gate and dam controlled drainage. Still, most exhibit buildings rested on piles driven deep into subsurface sand. Installation of underground utilities and landscaping, including the importation of ten thousand trees, completed the site preparation. Although the fair itself seemed, as the *Official Guide Book* claimed, "a vast city which progressed from blue prints to reality in an amazingly short time," the real "miracle" was "the largest single reclamation project ever undertaken in the eastern United States."[3] Visions of the future projected in the fair's exhibits by industrial designers collectively made only a footnote to the more solid career of urban planner and builder Robert Moses.

This subtlety was lost on thousands of people who converged daily on the fair by subway, train, bus, taxi, and private auto. Wherever they entered, they looked first for the spare, precise, stark white forms of the "theme center" (figures 134 and 135) looming above the fair's low architecture. The Trylon, a three-sided needle extending seven hundred feet into the air to a sharp point, symbolized, according to the *Official Guide*, "the Fair's lofty purpose."[4] Next to it, floating on barely visible piles over a circular pool, was the Perisphere, a smooth globe two hundred feet in diameter. A short arched bridge joined it to the Trylon at about sixty feet above the ground. From that point the Helicline, a wide flat ramp supported by slender round columns, spiraled down around both forms in almost a full circle to the ground. Up close, the stuccoed surfaces of the Trylon and Perisphere looked lumpy, as if someone had pasted burlap to wooden forms, but that unfortunate effect vanished at a slight distance. Abstracted, rendered in silhouette, the widely reproduced sphere and triangular needle became evocative images. Designed by architects Wallace K. Harrison and J. André Fouilhoux, they embodied the contradictory associations of the more ordinary streamlined architecture around them—the Trylon representing limitless flight into the future, the Perisphere controlled stasis.

The static vision gained expression inside the limiting curve of the Perisphere itself, where the fair's theme exhibit, designed by Dreyfuss, was mounted. After riding an escalator from the base of the Trylon into the Perisphere, visitors stepped onto one of two platforms, one above the other, revolving in opposite directions around the interior circumference of the globe. Apparently floating in space, they saw above them images of cloud and sky cast on the dome by concealed projectors and below them Democracity (figure 136), a meticulous model of "a perfectly integrated, futuristic metropolis" stretching fifteen scale miles to the horizon.[5] Although Dreyfuss referred to urban studies in devising this model, he intended Democracity not as "the perfect city plan" but as "a symbol of all city planning."[6] In his decentralized scheme, adapted from Ebenezer Howard's Garden City concept, a central city about two miles in diameter contained business, cultural, and leisure activities in low

135. Trylon and Perisphere. Architectural Forum.

*136. Dreyfuss. Democracity model inside the
Perisphere.* Architectural Forum.

buildings separated by generous green
spaces and connected by pedestrian walkways over streets and highways (figure 137).
Zoned according to use, this central area
ended abruptly on a river and a semicircular
beltway. Beyond stretched a greenbelt of
parks and farmland, dotted with twenty-five
satellite towns, some purely residential and
others residential-industrial, each with its
school and small business district but dependent on the center for all but necessities.

In this scheme, according to Dreyfuss,
leisure time would "not be dissipated in
idleness or carousing, but . . . employed in
improving man, physically and mentally . . .
through organized athletics, lectures, con-
certs and the direction of his pursuit of his
own hobbies."[7] A hint of the Ethical Culture
vision is here, but Democracity also incorporated ideas common throughout the fair.
The concept of rigid control, embodied in
the fair's thematic zones, reflected paternal-

istic faith in total environmental planning to
provide the good life for all. Fully planned in
every detail, Democracity offered a vision of
social stasis. Within its regulated zones, as
within the Perisphere's enclosing globe,
there was no room for growth. And it was
not, according to a variant of frequently
made observations, the "impossible dream of
a Jules Verne or an H. G. Wells," but a city
that "could be constructed tomorrow with
the technological knowledge we [now]
have."[8] In this sense Democracity reflected
most exhibits at the fair, which was described by a critic as in general "looking at
Tomorrow and arriving mostly in the middle
of Today." By displaying contemporary technologies and industrial processes in futuristic
architectural settings, commercial exhibitors
implicitly stated that the future was already
here if people would only realize it.[9]

The techniques used by Dreyfuss to present Democracity to fair-goers echoed those

137. Dreyfuss. Democracity. Dreyfuss Archive.

*138. Overview of New York World's Fair of 1939
with Transportation Zone in foreground.*
Architectural Record.

employed throughout the fair. Rapid movement of large crowds required devices like
escalators and revolving platforms. But these
tended to make people passive consumers of
simplistic images displayed quickly and
dramatically. Intricate subtleties of Dreyfuss's plan were lost on most viewers, who
became entranced by an accompanying
orchestral and choral composition, a vague
narration "stressing a lot of cosmic points
about humanity,"[10] and a grand finale to the
six-minute show in which projected images
of groups of workers representing various
trades marched forward from a distance,
finally forming a gigantic ring of inter-
dependent humanity around the interior
dome of the Perisphere. Harmless enough in
the theme center, streamlined circulation
techniques and futuristic display mechanisms gave manufacturers the means for im-
pressing their own vision of the future on the
minds of consumers. These methods implied
the possibility of an imperceptibly controlled future society far different from
Dreyfuss's pastoral Democracity.[11]

One observer reported that the theme
center's "dramatic punch" is felt "when you
step off the moving platform and find your-
self suddenly outdoors, on the Helicline, suspended high above the fountains playing on
the Perisphere, with the living diorama of
the Fair itself below you—with all its cock-
eyed lights, colors and angles—and the
ramp, black with slow moving figures,
sweeping around the great ball to terra firma
—a stage set for some fantastic, futuristic
dream."[12] Looking out from the ramp's first
downward sweep, a visitor saw the Transportation Zone (figure 138), appropriately
separated from the rest of the fair by the
Grand Central Parkway. Map in hand, one
could pick out the Aviation and Marine
Transportation buildings' vaguely representational forms, Teague's gear-shaped Ford
building with a spiraling outdoor auto test
ramp, and Geddes's flowing, free-form
General Motors building. Beyond, separated
by a semicircular tree-lined promenade, lay

139. Loewy. S-1 locomotive for Pennsylvania Railroad, 1939. Raymond Loewy.

the conventionally streamlined forms of Loewy's Chrysler building, its ovoid center section flanked by slender pylons with streamlined chevrons, and the Eastern Railroads building. There one could later view the world's largest steam locomotive, Loewy's bullet-nosed Pennsylvania S-1 (figure 139), driving furiously nowhere on a treadmill.[13]

Winding clockwise down the Helicline, one touched ground on the Perisphere's opposite side, from which the fair's main axis extended northeast as a wide, tree-lined mall, broken by rectangular reflecting pools, a mammoth statue of Washington, a circular fountain-filled Lagoon of Nations—terminating in a hopefully named Court of Peace, with the Federal building at its head, looking like an import from Mussolini's Italy. Two diagonal avenues radiating north and east from the Perisphere contributed to this formal layout, which architectural critic Talbot

F. Hamlin found incongruent to the "rococo" (that is, streamlined) forms of the fair's buildings.[14] Stretching from the Perisphere to the lagoon were exhibits of business and industry, divided into zones of Communications, Community Interests, Food, and Production and Distribution. Beyond lay exhibits of individual states and foreign nations.

To the southeast, separated from the fair's more serious aspects by Horace Harding Boulevard, was the Amusement Zone. There one could stroll along Fountain Lake, relax in half-timbered Merrie England, watch swimmers' patriotic precision drills in Billy Rose's Aquacade, or enjoy a nightly sound-light-fountains-fireworks extravaganza over the lake. Or, for a mere fifteen cents, one could enter Geddes's Crystal Lassies concession, a "polyscopic paradise for peeping Toms" with "enough angles to confuse an Einstein." Looking through one-way mirrored glass, viewers on three levels watched a single topless dancer on a twelve-foot-square glass platform, her image multiplied hundreds of times by the octagonal structure's multi-faceted mirrors.[15]

As a whole, the fair lacked the illusion of precision provided by Geddes's show. Its halfhearted formal layout failed to pull together a riotous confusion of architectural forms whose only common denominators were streamlines, pylons, bright reds and yellows, and shiny metal trim. Unlike Democracity, the unified conception of a single intelligence, the fair as a whole was the product of hundreds of competing firms, each seeking to attract attention. As Loewy frankly admitted, in an apt metaphor for consumer-oriented machine-age America, it was "a huge department store."[16]

According to Teague, the fair provided manufacturers "an opportunity to state the case for the democratic system of individual enterprise" at a time when other nations were adopting collectivism. By collaborating with industry on this project, industrial design was thus "emerging for the first time in its major, basic role—as the interpreter of industry to the public."[17] As Teague seemed to admit, however, participation in the fair subjected designers to the art-and-industry dilemma that had plagued them throughout the decade: did promotion of sales naturally lead to beauty? Teague thought so. He stated in 1937 that although the world's fair would function primarily as "a place where merchants come to display their wares to possible purchasers," it would also be "aesthetically beautiful—a vast, magnificent work of art."[18]

Much of the fair's own publicity, which sounded as if Teague had written it, revealed a split between the ideal and the real. An expensive promotional volume issued in 1936 contained a visionary foreword by fair president Grover Whalen. With an implied nod to H. G. Wells, Whalen suggested that the fair would predict and possibly dictate the "shape of things to come" by exhibiting "the most promising developments of ideas, products, services and social factors of the present day in such a fashion that the visitor may, in the midst of a rich and colorful festival, gain a vision of what he might attain for himself and for his community by intelligent, coöperative planning toward the better life of the future."[19] Whalen more clearly spelled out his commercial meaning the following year in a public relations pamphlet. Not nearly as visionary as before, Whalen declared that "business and industry possess *today* most of the implements and materials necessary to fabricate a new World of Tomorrow." Society did not need "new inventions and new products" but "new and improved ways of utilizing existing inventions and existing products." The fair, according to Whalen, would help "sell a new method or procedure or way of life involving the use of many new [*sic*] products" by bringing "large groups of producers and distributors ... into direct, planned and simultaneous contact with great masses of consumers." The fair thus would provide business and industry with "an opportunity to construct their *own* World of Tomorrow, to build for the future along lines patterned and planned." In other words, the fair would embody "the specifications" for a consumer society planned not cooperatively by its citizens but by producers and distributors who hoped to promote sales of individual products to individual citizens.[20]

Visitors to the fair found the popular image of planning embodied in Democracity, in Teague's diorama for U.S. Steel, showing a Corbusian city whose towers were separated not by parks but by a labyrinth of multilevel highways, and even in Loewy's Wellsian "rocketport of the future," an animated diorama (figure 140) depicting a rocket gun capable of shooting a projectile full of passengers from New York to London in one hour. But the fair's true vision of the future was presented in the Town of Tomorrow, a group of predominantly traditional homes furnished with a wide variety of conveniences and appliances. Consumer en-

140. Loewy (right) supervising construction
of the rocketport of the future for the
transportation focal exhibit. Raymond Loewy.

gineering constituted the fair's major element of social planning.

The arrangement of fair buildings was
chaotic, but individual structures were carefully planned to instill specific images in the
minds of visitors. With typical perceptive-
ness Douglas Haskell observed that the fair
witnessed the emergence of "architecture as
environmental control" rather than "the
mere enclosing of space." Nineteenth-
century exhibition buildings were vast sheds
meant only to shelter typical objects,
machines, and cultural artifacts of civilization. By 1939, however, attention had shifted
from objects to their potential consumers.
By "taking care of the visitor's every requirement, actual or fancied, not to mention

his guidance," exhibition buildings had become machines for processing people.[21]
Unlike former expositions, which often
brought people up sharp in amazement,
making them stop to ponder the mute significance of the Corliss engine or Edison's
light bulb, the streamlined 1939 fair eliminated thought-provoking disturbances in
order to present superficial images. Teague,
who so often articulated what designers
were doing, insisted that "people must *flow*
in an exhibit." They had to "follow the line
of least resistance just as water does."[22]
Exhibits had to be engineered "in such a way
that the spectator's interest is stimulated and
his responses are involuntary."[23]

Fatigue was the major obstacle to making

an impression on fair visitors, according to Teague. Reeling under *the impact of many wholly unrelated impressions on the consciousness,*" the average person experienced a lowering of his "mental receptivity" to that of a twelve-year-old.[24] Teague presented this observation as a challenge to designers, but fatigue probably rendered people more passive, more open to suggestion, than they normally were. Whatever the case, he insisted that to present his "message" an exhibitor must dramatize it "in visual form so that the average visitor cannot escape its import."[25] An exhibit had to relate simply and obviously to its viewers' everyday lives. It had to be brightly colored and full of motion to attract and hold attention. Finally, sections of an exhibit had to be arranged in a "logical order" with a "dramatic structure" leading to a climax. An audience did not leisurely explore an exhibit; it was "moved past the acts." Controlled circulation, both symbolized and expedited by streamlined architecture's flowing interior walls, summarized Teague's philosophy of exhibition design (see figure 141). As a behaviorist, he insisted that after "we lead them [visitors] craftily through a planned maze we should reward them at the end with a chance to rest and relax."[26]

With this emphasis on movement and drama, common throughout the fair, exhibits attracted more attention to themselves than to the machines, technologies, or processes they represented. The Borden Company won acclaim for a demonstration of mechanical milking carried out for no logical reason on a revolving platform called a "rotolactor." Sound reproduction on magnetized steel tape received attention in AT&T's building. Visitors conversed with one another in chairs on a stage covered with tulips. Their brief talk ended, the stage swung out of sight to be replaced by another on which sat dummies who repeated the conversations. More dramatic was Du Pont's demonstration of insecticides. From thousands of houseflies a technician chose a few to be introduced

"into a glass-walled death chamber" and exterminated "by an atomized spray." Refrigeration, inherently motionless, taxed designers' imaginations. Chrysler's Airtemp division advertised itself with a "frozen forest" of life-size tropical plant forms covered with a thick layer of frost generated by imbedded refrigeration coils. Whenever possible, animated models replaced photographs and charts. This technique reached its apex in Teague's "Ford Cycle of Production" (figure 142) praised by *The Architectural Forum* as "the most impressive display at the fair." A series of concentric steps rose thirty feet above a circular platform a hundred feet in diameter. Eighty-seven mechanical puppets represented steps in the manufacture of an automobile from raw materials, at the bottom, to finished product, three of which perched above on a smaller circular platform. Reminiscent of a wedding cake, the entire 152-ton structure rotated.[27]

Some visitors no doubt learned something about auto manufacturing from this exhibit. Others took away the message that Ford's operations benefited thousands of people in seemingly unrelated industries. But most, too tired to absorb information, delighted in the mechanics and scope of the display itself. Visitors would soon forget even diluted details of a particular manufacturing process, but they would remember the ingenuity of its demonstration or explanation. Each display functioned as an advertisement intended to leave vague impressions of a corporation's enterprise and public beneficence.

An exposition no longer served as a museum of contemporary civilization but as a vast three-dimensional package for the consumerist way of life. Freed from considering any function other than lubricating the production-distribution machine, industrial designers could concentrate on consumer engineering to create public demand for a future society that would give most benefit to private corporations. This approach considered the average person only as a con-

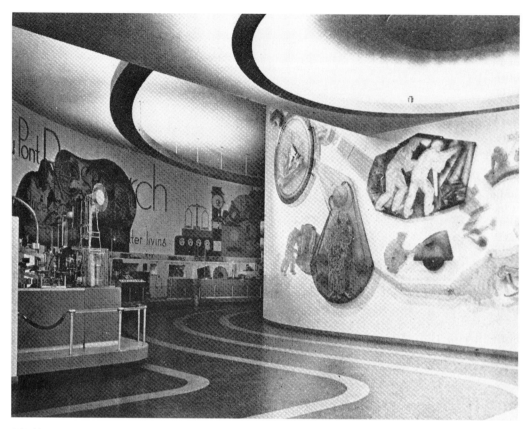

141. Teague. Interior of Du Pont exhibit.
Architectural Forum.

sumer, a passive individual receiving im-
pulses, prodded, stimulated, and living
packaged experiences. The 1939 fair held
no projections, no bottlenecks to jog the
visitor's mind into activity. Everything, from
architecture to exhibits, was streamlined,
rounded off, and reduced to the lowest level
of public comprehension. The fair's official
publications revealed this lack of content.
Earlier expositions, most notably the Columbian, but to an extent the Century of Progress as well, generated serious volumes on
the arts, religion, society, and technology.
Except for an art catalogue edited by Holger
Cahill, the New York fair produced only
souvenir booklets and reams of publicity
material issued in praise of exhibits.[28] The
fair's verbiage flowed as smoothly and efficiently as the images in its exhibits and the

people who viewed them. The fair reflected
worship of frictionless processing, the lone
individual becoming as irrelevant as a single
automobile.

The General Motors building (figure 143)
with its Futurama, one of the few exhibits
to glorify planning as the public understood
it, succeeded more than most in directing
the flow of visitors.[29] Architecture critic
Haskell, usually reserved, became "ecstatic
over the strange power of the streamlined
complex of General Motors, so like some
vast carburetor, sucking in the crowd by
fascination into its feeding tubes, carrying
the people through the prescribed route,
and finally whirling them out, at the very
center of the display, so they might drift
outward in free dispersion."[30] As visitors
crossed to the Transportation Zone by the

142. Teague. Cycle of Production, Ford building.
Architectural Forum.

northern bridge over the Grand Central Parkway, they passed on their left a sharply curved corner of the GM building, its flowing forms "a translation of the streamline forms" of auto designs, its surface painted a metallic silver-gray "to simulate Duco finish on automobiles."[31] The building's long horizontal wall rose gradually along its length, doubling back to end in a high concave screen, which blocked a view of the rest of the Zone and visually hooked visitors onto either of two slender ramps. These ramps, after meandering beside the building, vanished into a narrow slit in the facade. Geddes and his asso-

ciates, who included architect Eero Saarinen, planned this side of the building solely to "catch" the "stream of pedestrians" from the bridge.[32]

Those caught, some 27,500 daily, entered a "haunting blue-black" auditorium.[33] As they went down switchback ramps to the front, a recorded voice backed by soothing music directed attention to a huge map of the United States, which electronically compared traffic volume on major highways in 1939 with that predicted for 1960 and then depicted a solution to the mess—a superhighway system. At the front of the audi-

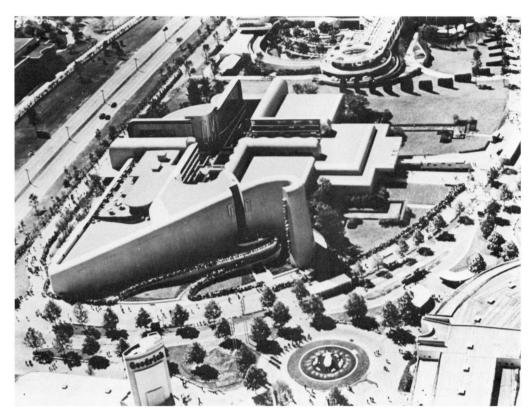

143. *Geddes. General Motors building. Geddes Collection.*

torium uniformed ushers directed visitors to a row of plush, high-backed seats, "so designed as to suggest a private, traveling opera box."[34] As the rubber-tired train—two seats to a car, twelve cars to a train—pulled away from the loading platform, a voice synchronized for each car invited passengers to enjoy a fifteen-minute simulated airplane ride over the America of 1960. Visitors traveled some 1,600 feet, looking through a continuous transparent screen over vast landscape dioramas covering 36,000 square feet on three levels and containing a million trees, half a million buildings, and fifty thousand streamlined automobiles, ten thousand of them in motion.[35]

For the most part, with two essential exceptions, the projected America of 1960 resembled that of 1939. Such innovations as a scientific orchard with trees under glass, a

giant power dam, and a futuristic amusement park hardly intruded on grand landscapes of farmland, mountains, forests, and rivers. This predictable, if spectacular, background highlighted all the more two major departures from the realities of 1939—a superhighway system and a Corbusian metropolis. Geddes led his captive audience along the route of a monumental fourteen-lane highway, divided in each direction into two fifty-mile lanes, two seventy-five-mile lanes, and a single hundred-mile lane. Grass strips and physical barriers separated individual lanes, joined periodically by acceleration and deceleration lanes, while, when terrain permitted, several miles of land separated the two directions of travel. Stationed in periodic bridges over the highway, according to the narrator, were technicians who monitored traffic flow and radioed directions to motorists for

144. Geddes. Highway intersection in the
Futurama. Geddes Collection.

entering the highway and changing lanes. A
night sequence revealed that headlights
were unnecessary because each lane was lit
by continuous tube lighting installed in side
barriers. At an intersection with another
superhighway Geddes replaced the ten-year-
old cloverleaf with a maze of ramps making
possible a fifty-mile-an-hour ninety-degree
turn in either perpendicular direction
(figure 144).

As the highway approached a metropolis
patterned on St. Louis, feeder highways
doubled its number of lanes, and it crossed
a four-level bridge over the river without
bottlenecks. Following an approach highway,
armchair futurists reached "an American
city replanned around a highly developed
modern traffic system," as the narrator
described it (figure 145). Widely spaced,
quarter-mile-high streamlined skyscrapers

separated by parks and lower buildings predominated. Every ten blocks an elevated
expressway with lower access roads cut
across the city, while all sidewalks were
raised above street level and bridged four
ways at each intersection. After a close-up
view of an intersection (figure 146), visitors
emerged from the Futurama to find themselves standing outside in that very intersection (figure 147), streets below filled not
with teardrop autos but with current GM
models.

Geddes's Futurama proved to be the hit of
the fair, prompting an observer to note that
it "combines the thrills of Coney Island with
the glories of Le Corbusier."[36] Despite the
entertainment value of the GM extravaganza,
with which the corporation hoped to reap a
good return in publicity and auto purchases
on an investment of $8 million, Geddes took

*145. Armchair visitors to the Futurama's city of
1960. Geddes Collection.*

146. Model street intersection in the Futurama.
Geddes Collection.

*147. Full-size street intersection in the GM
building. Geddes Collection.*

seriously his vision of 1960. Many of its ideas
were derivative. William Stott has pointed
out that Geddes borrowed the major outlines of his city, as well as a circular airport, from Le Corbusier. In addition, the
designer took his pedestrian walkways in
every detail from a proposal made by his
friend Harvey Wiley Corbett to the New
York Regional Plan Association in 1923. And
his suggestion of built-in highway lighting
temporarily activated by passage of a car past
photoelectric cells, an idea too complex for
dramatization, came from Raymond Loewy.[37]
But Geddes since 1931 had been looking for
solutions to the problems of urban traffic
congestion and mounting deaths on open
highways.[38] Late in 1936, with the recommendation of Stanley Resor of the J. Walter

Thompson Agency, Geddes signed a contract
with Shell Oil to provide models illustrating
rural and urban traffic solutions for an advertising campaign. This project introduced him
to Miller McClintock, director of Harvard's
Studebaker-funded Erskine Bureau for Street
Traffic Research.

McClintock, who earned a Harvard doc-
torate in 1924 with a dissertation on traffic
control, was described by *Fortune* in 1936
as the "No. 1 man" in the field.[39] His "friction
theory" of traffic coalesced with industrial
design's predilection for streamlining, for
eliminating friction in all areas of life. All
automobile accidents and traffic congestion,
he found, resulted from one of four frictions:
medial, the friction of opposing directions of
travel; marginal, the friction of autos with

parked cars, pedestrians, and fixed objects; intersectional; and internal-stream, the friction of automobiles passing others in the same direction of travel, which McClintock found responsible for 47 percent of all accidents. Pedestrians, however, composed 65 percent of urban traffic fatalities.[40] From these facts came the two solutions of limited-access highways and separation of pedestrians from vehicular traffic. Geddes, however, proposed a novel solution for eliminating urban congestion. In staff meetings dedicated to the Shell project he repeatedly suggested banning private automobiles and constructing large parking garages on urban fringes, from which commuters would be ferried downtown by minibuses. McClintock, the expert consultant, scotched the idea because people could not be bothered with the irritation of switching to another mode of transportation. In the future the city would become a frictionless traffic machine.[41]

Geddes's Shell project yielded a six-foot triangular model (figure 148) of a "city of tomorrow" that could have been inserted into the Futurama. Shell used photographs of the model in the planned advertising campaign and made a film available to service clubs and chambers of commerce to gain support for a national highway system.[42] Using slides of the model and of sketches of the superhighway system that he and Geddes had devised, McClintock addressed the National Planning Conference in Detroit on the subject.[43] Before Shell's advertisements ceased running, Geddes himself envisioned a world's fair exhibit to dramatize McClintock's theories.[44] When in May 1938 General Motors abandoned architect Albert Kahn's completed plans in order to sign a contract with Geddes, he had already completed preliminary plans for the entire building and had tried unsuccessfully to sell them to Shell and Goodyear.[45] Geddes's Futurama went well with GM's long-term goals. President Alfred P. Sloan, Jr., had in 1932 organized a National Highway Users Conference, incor-

148. Geddes. Shell Oil's "city of tomorrow," 1937. Geddes Collection.

porating elements of the automobile, petroleum, and rubber industries, whose purposes included lobbying in Washington for a federally funded highway system, whose convenience would boost automobile sales.[46] And Geddes's plans for streamlining the flow of urban automobile traffic appealed to a corporation that for several years had bought up streetcar lines, scrapped their equipment, introduced buses, and then sold out, often with a proviso that purchasers not return to rail transit.[47] The Futurama's metropolis of the future contained no mass transportation other than buses.

General Motors officials hoped the Futurama would arouse public support for a highway system of some sort, whether or not it exactly resembled the one in their diorama. But Geddes, enthralled by his project's scope—monumental on its own scale—hoped to see his vision actualized. In the planning stage his staff discussed details as if they really were designing highways and reconstructing cities. Members of a "traffic bureau" spent hours making certain that every single intersection was "absolutely practical," but this degree of realism was lost on fair-goers gliding by in their upholstered chairs.[48] Geddes himself incorporated the smallest details in a perspective he had learned twenty-five years before, perched on the highest catwalk of Detroit's Garrick Theatre, an experience that had led him as a

149. Geddes placing buildings in the city of tomorrow. Geddes Collection.

stage designer to emphasize "organizing all phases of stage activity so that they move in a series of traffic patterns, so planned that they cannot collide or interfere with each other."[49] In a letter to his wife he recounted the "funny experience" of arranging Futurama buildings on a forty-by-sixty-foot map of St. Louis. He and half a dozen staffers "walked around with pockets and hands full of skyscrapers," arranging them for a "whole effect" from a bird's-eye view. She wished she had seen him "playing God . . . plunking down skyscrapers where you want and spending millions as you choose" (see figure 149).[50] Geddes described this scene with whimsy, but he considered the Futurama a blueprint he hoped to see used.

During early negotiations with GM, Geddes suggested presenting the Futurama "as the result of an independent research [project]" to remove from it "the curse of the advertising angle" and render its principles attractive to "outside agencies such as the Government."[51] Shortly before the fair's opening, Geddes arranged a dinner invitation to the White House. Roosevelt listened politely to Geddes's proposals for a national highway system, but he seemed more interested in a streamlined yacht model the designer brought along.[52] Geddes also tried to recruit Robert Moses, the one man capable of realizing his urban traffic solution, who had refused earlier to look at the Shell

model.[53] Moses prompted an exchange of views by describing Geddes's plans as "bunk" at an annual meeting of the American Society of Civil Engineers. The nation did not need mammoth cross-country highways. It needed urban expressways to make internal traffic flow more smoothly.[54] Geddes replied with a press release carried in part by the *New York Times*. Not only, he argued, did Moses show lack of foresight in failing to plan for the cross-country traffic of 1960, but even his own New York expressways and bridges became clogged the moment they opened. In his statement Geddes admitted "great admiration" for Moses as "Park Commissioner." In fact, Geddes continued, "no one excels him in resetting bushes in a park and finding play spaces for the underprivileged voters," but "landscaping along a highway does not make it safe for present-day travel."[55]

With this sarcastic remark Geddes curiously reversed roles with Moses. New York City's highway czar became a mere decorator, a stylist of highways unconcerned with function, while Geddes, industrial designer and creator of the Futurama highway system, became the serious planner and builder. Over the following months the two men, who were apparently on amiable terms personally, exchanged a series of letters growing out of the brief public controversy. Geddes continued to insist on his work as an accurate blueprint for actual construction, while Moses regarded the Futurama "as work of the imagination." From Geddes he sought admission "that you don't pose as an expert . . . that you are simply taking a look into the future, that you don't want your recommendations to be taken too seriously by practical work-a-day people who must live in their own time." By admitting so, Geddes would be "paving the way for further recognition as a creative artist."[56] Several months later Moses himself paid tribute to Geddes's aesthetic gifts by awarding him a contract for designing playground equipment.[57] Thus ended Geddes's last attempt to gain support from individuals whose power equalled the scope of his reconstruction project. Undaunted, he ended his Futurama association with a gesture of pure surrealism. Discussing with GM executives the exhibit's fate at the close of the fair's second year, he suggested flying around the country with the entire Futurama in a dirigible.[58]

The story of the Futurama, especially Geddes's extravagant hopes and their eventual frustration, paralleled industrial design's fate in the thirties and beyond. Whatever his proposals' worth, Geddes expressed a genuine concern for reconstructing the American environment on a grand scale that extended back through Teague to Calkins, the first publicist of industrial design. In idealistic moments designers believed that their method, which involved reducing friction or removing irritations, would contribute to making life more efficient and thus potentially more fulfilling. From the simplest product to the most complex metropolis, all artifacts of machine-age America would reflect their vision. And the streamlined style would accurately express the function of reducing friction and would provide America a visual coherence never before attained. The world of the future, foreshadowed by the New York World's Fair of 1939, would embody a designed stability unknown during the uncertain Depression years.

This common vision of industrial design's larger cultural role was never realized. Not even streamlining, its stylistic embodiment, which in 1939 had seemed the wave of the future, survived World War II intact. In the fifties the austere lines of the International Style dominated American architecture, while "streamlining" referred mostly to Detroit's tailfins and chrome. The war itself contributed to the vision's failure. A nation exhausted by fifteen years of deprivation and war naturally embraced a resurgence of the commercial tempo previously experienced in the twenties. Novelty and constant change seemed more desirable than

a stasis that could hardly sustain prosperity. Even more important, the monolithic coherence of industrial design's machine-age world—and its bent for processing people through curved flow lines—recalled European and Russian totalitarianism. As America entered the Cold War, social planning, a prime concern of society during the Depression, became ideologically suspect. The free enterprise system, certainly as much at odds with design coherence as with social control, became an unquestioned element of the faith of loyal postwar Americans.

Even during the thirties, the American economy lacked the directed unity of purpose required for realization of industrial design's vision. A designer like Geddes could receive $8 million to create a model of his planned future, but he had no leverage to ensure its adoption by government. His patron, General Motors, did not regard the project as an actual blueprint but as an exercise in consumer engineering useful for influencing public opinion. The Futurama did contain intricate realistic detail, but its presentation revealed its superficial nature. The designer himself recognized the limitations of the Futurama, which accompanied a brief ride with a vague recorded narration, when he followed it with *Magic Motorways*, a book that described his plans in detail.[59] Industrial designers, despite their idealistic intentions, could not rise above the designs of their employers. After the war they did not often try to.

Industrial design eventually became institutionalized as a sales technique comparable to advertising. Naive images of social control, like streamlined flow lines, disappeared for the most part, but designers refined their techniques of manipulation in the service of individual corporations with little thought for the total effect. Coherence disappeared as a goal, at least on any scale larger than shopping malls or computerized amusement parks—heirs of the techniques of social control developed for the World's Fair. When designers "returned from the war," for which they had provided camouflage techniques, artillery housings, training manuals, propaganda displays, and a war room for the Joint Chiefs, they concentrated on fueling the flow of goods of the affluent postwar period. An occasional designer continued to express small-scale idealism in an individual product, but the Depression dream of building a harmonious machine-age America dissolved. The pressure of artificial obsolescence, always a major undercurrent in the thirties, finally won out. The streamlined style, discredited owing to hints of totalitarianism in its smooth-flowing lines and its ideal of permanence, yielded to a flood of extravagant forms whose impermanence and lack of coherence could not be dignified by the word "style." More recent industrial designers have indeed shaped "the world of tomorrow"—its appliances, franchise architecture, and eternally "new" corporate logotypes—but it is not the world envisioned in their best moments by the founders of the profession.

Prologue

1. See Warren I. Susman, "The Thirties," in *The Development of an American Culture*, ed. Stanley Coben and Lorman Ratner (Englewood Cliffs, N.J.: Prentice-Hall, 1970), pp. 179–218; Alfred Haworth Jones, "The Search for a Usable American Past in the New Deal Era," *American Quarterly* 23 (Dec. 1971), 710–24; William Stott, *Documentary Expression and Thirties America* (New York: Oxford University Press, 1973); and Richard H. Pells, *Radical Visions and American Dreams: Culture and Social Thought in the Depression Years* (New York: Harper and Row, 1973), particularly chapters "Documentaries, Fiction, and the Depression" and "The Decline of Radicalism, 1935–1939."
2. Steinbeck (New York: Viking, 1939), p. 176.
3. Robert S. Lynd and Helen Merrell Lynd, *Middletown in Transition: A Study in Cultural Conflicts* (New York: Harcourt, Brace, 1937), p. 469.
4. The speech, delivered Nov. 1, 1933, is in the Egmont Arens Collection, George Arents Research Library for Special Collections at Syracuse University (hereafter referred to as *Arens*), box 51. The A&P Carnival is described in *Official Guide Book of the Fair* (Chicago: A Century of Progress, 1933), p. 103.
5. Lynd, *Middletown*, p. 247.
6. Clipping, *Arens*, box 59.
7. Giedion, *Mechanization Takes Command: A Contribution to Anonymous History* (1948; rpt., New York: W. W. Norton, 1969), p. 3.

1. A Consumer Society and Its Discontents

1. Giedion, *Mechanization Takes Command: A Contribution to Anonymous History* (1948; rpt., New York: W. W. Norton, 1969), p. 121.
2. Robert S. Lynd with Alice C. Hanson, "The People as Consumers," in *Recent Social Trends in the United States: Report of the President's Research Committee on Social Trends* (New York: McGraw-Hill, 1933), p.

857. See also William Leuchtenburg, *The Perils of Prosperity 1914–32* (Chicago: University of Chicago Press, 1958), p. 194; Thomas C. Cochran, *American Business in the Twentieth Century* (Cambridge: Harvard University Press, 1972), p. 31; James J. Flink, "Three Stages of American Automobile Consciousness," *American Quarterly* 24 (Oct. 1972), 462–63; and Gilman M. Ostrander, *American Civilization in the First Machine Age: 1890–1940* (New York: Harper Torchbooks, 1972), pp. 9–10.
3. See Leuchtenburg, pp. 178–79; George Soule, *Prosperity Decade: From War to Depression: 1917–1929* (New York: Rinehart, 1947), pp. 127–31. On factory management see Samuel Haber, *Efficiency and Uplift: Scientific Management in the Progressive Era 1890–1920* (Chicago: University of Chicago Press, 1964).
4. U.S., Bureau of the Census, *Historical Statistics of the United States: Colonial Times to 1957* (Washington, D.C.: Government Printing Office, 1960), pp. 462, 14.
5. See Douglas A. Fisher, *Steel Serves the Nation 1901–1951: The Fifty Year Story of United States Steel* (New York: U.S. Steel Corporation, 1951), p. 41; William S. Dutton, *Du Pont: One Hundred and Forty Years* (New York: Scribner, 1951), pp. 295–301.
6. Lynd and Hanson, "People as Consumers," pp. 866–67. See also Daniel J. Boorstin, *The Americans: The Democratic Experience* (New York: Random House, 1973), pp. 422–28.
7. See comments on class distinctions, ibid., pp. 860–61. The next two paragraphs are based on charts on pp. 896–901.
8. *Historical Statistics*, p. 417.
9. Boris Emmet and John E. Jeuck, *Catalogues and Counters: A History of Sears, Roebuck and Company* (Chicago: University of Chicago Press, 1950), pp. 338–57.
10. Malcolm M. Willey and Stuart A. Rice, "The Agencies of Communication," in *Recent Social Trends*, p. 216. For statistics see Lynd and Hanson, "People as Consumers," p. 909; *Historical Statistics*, p. 491.
11. Lynd and Hanson, "People as Consumers," p. 874.

12. Poffenberger, *Psychology in Advertising* (Chicago and New York: A. W. Shaw, 1925), pp. 28–29. See Otis Pease, *The Responsibilities of American Advertising: Private Control and Public Influence, 1920–1940* (New Haven: Yale University Press, 1958), p. 34; Merle Curti, "The Changing Concept of 'Human Nature' in the Literature of American Advertising," *Business History Review* 41 (Winter 1967), 335–57.

13. Calkins, *Business the Civilizer* (Boston: Little, Brown, 1928), pp. 24–25, 15.

14. Willey and Rice, "Agencies of Communication," p. 217.

15. Updegraff, *The New American Tempo: And the Stream of Life* (Chicago: A. W. Shaw, 1929), pp. 2–3, 6, 8–11. For similar observations see John Vassos and Ruth Vassos, *Contempo: This American Tempo* (New York: E. P. Dutton, 1929).

16. Updegraff, *The New American Tempo*, p. 15.

17. Ibid., pp. 36–37.

18. See his autobiography, *"And Hearing Not—": Annals of an Adman* (New York: Scribner, 1946).

19. Calkins, *Modern Advertising* (New York: D. Appleton, 1905), pp. 3–4. Holden's name appeared on the title page as a courtesy.

20. *"And Hearing Not—"*, p. 239.

21. *Business the Civilizer*, p. 22.

22. Calkins described the campaign and reproduced an advertisement and promotional brochure from it in *The Business of Advertising* (New York: D. Appleton, 1915), pp. 205–209.

23. See his autobiography, *So Away I Went!* (Indianapolis: Bobbs-Merrill, 1951).

24. "Art and the Motor Car," *Bulletin of the Society of Automotive Engineers*, Sept. 1916, pp. 739, 743.

25. Norman G. Shidle, "'Beauty Doctors' Take a Hand in Automotive Design," *Automotive Industries*, Aug. 13, 1927, p. 218.

26. Soule, *Prosperity Decade*, pp. 276, 285–88.

27. As quoted in "Color in Industry," *Fortune* 1 (Feb. 1930), 90.

28. Alfred P. Sloan, *My Years with General Motors*, ed. John McDonald with Catharine Stevens (Garden City, N.Y.: Doubleday, 1964), p. 269.

29. As quoted in ibid., p. 268.

30. As quoted in ibid., p. 272; and by Boorstin, *The Americans*, p. 552. Boorstin's treatment of the annual model is instructive.

31. As quoted by Philip Van Doren Stern, *Tin Lizzie: The Story of the Fabulous Model T Ford* (New York: Simon and Schuster, 1955), p. 61.

32. The cost estimate is from Allan Nevins and Frank Ernest Hill, *Ford: Expansion and Challenge 1915–1933* (New York: Scribner, 1957), p. 452. They cover the episode on pp. 379–478.

33. As quoted in ibid., p. 455.

34. As quoted by Frederick P. Keppel and R. L. Duffus, *The Arts in American Life* (New York: McGraw-Hill, 1933), pp. 133–34. If apocryphal, these statements become more significant.

35. Paul Thomas, "The Secret of Fashion and Art Appeal in the Automobile," *Society of Automotive Engineers Journal* 23 (Dec. 1928), 595–96.

36. Frederick P. Keppel, "The Arts in Social Life," in *Recent Social Trends*, p. 977.

37. Ralph Abercrombie, *The Renaissance of Art in American Business*, General Management Series, no. 99 (New York: American Management Association, 1929), pp. 6–7.

38. Paul Thomas, "Scientific Textile Styling," in *How the Manufacturer Copes with the Fashion, Style and Art Problem*, General Management Series, no. 98 (New York: American Management Association, 1929), p. 6.

39. Keppel and Duffus, *Arts in American Life*, p. 143.

40. Ibid., p. 207.

41. Abercrombie, *Renaissance of Art*, pp. 3–6.

42. See *The Use of Style and Design in Industry* (New York: Metropolitan Life Insurance Company, [1928]), pp. 5–6; Earnest Elmo Calkins, "Beauty the New Business Tool," *The Atlantic Monthly* 140 (Aug. 1927), 150.

43. "Color in Industry," *Fortune* 1 (Feb. 1930), 85; "Color in Electric Iron Handles," *Plastics and Molded Products* 4 (July 1928), 398.

44. See discussion by Stanley Needham in E. Grosvenor Plowman, *Fashion, Style and Art Spread to Other Lines of Business*, General Management Series, no. 106 (New York: American Management Association, 1929), p. 29.

45. Edward L. Bernays, *Biography of an Idea: Memoirs of Public Relations Counsel Edward L. Bernays* (New York: Simon and Schuster, 1965), p. 302.

46. For a typical discussion see Oswald W. Knauth, "The Effect of the Public's Demand for Better Art on the Technique of Merchandising," *Harvard Business Review* 7 (July 1929), 406–12.

47. "Saks is Very . . . ," *Fortune* 18 (Nov. 1938), 126.

48. De Forest, "Getting in Step with Beauty," *The American Review of Reviews* 77 (Jan. 1928), 66. On the Macy exhibit see "Cheap

and Smart," *Fortune* 1 (May 1930), 86.

49. Fairclough was commenting on Irwin D. Wolf's "Style Organization from the Executive's Viewpoint," in *How the Retailer Merchandises Present Day Fashion, Style and Art*, General Management Series, no. 97 (New York: American Management Association, 1929), p. 8.

50. Keppel, "The Arts in Social Life," p. 979.

51. "Vogue and Volume: Mass-Production in a New Quandary," *The J. Walter Thompson News Bulletin*, no. 136 (Nov. 1928), pp. 7–10.

52. Plowman, *Fashion, Style and Art*, p. 15.

53. The Creange formula first appeared in the report of the U.S. Commission, *International Exposition of Modern Decorative and Industrial Art in Paris, 1925* (Washington, D.C.: Government Printing Office, 1925), pp. 24–25.

54. Eventually, "stylist" became a term of reproach aimed at designers concerned only with surface appearances. To call someone a "stylist" was to impugn his motives, to deny him the integrity of the title "industrial designer." In a cassette letter to the author recorded Feb. 29, 1976, Raymond Loewy stated that "styling" was a word and a concept that had "plagued our profession and plagued me personally." Trained as an engineer, he was "against styling and for logical design."

55. Wolf, "Style Organization," pp. 4–5.

56. Alcott, "*Style* Challenges/Industry Answers," *Factory and Industrial Management* 77 (March 1929), 466–67.

57. "Cannon: I," *Fortune* 8 (Nov. 1933), 52.

58. Reported in "The Eyes Have It," *The Business Week*, Jan. 29, 1930, p. 30.

59. Title of an article by John Cotton Dana in *Forbes*, Aug. 1, 1928, pp. 16–18, 32.

60. Calkins, "Beauty the New Business Tool," *Atlantic*, Aug. 1927, pp. 156, 151, 150. The comparison to medieval Europe, here only implied, became explicit in "The New Consumption Engineer, and the Artist," in *A Philosophy of Production: A Symposium*, ed. J. George Frederick (New York: Business Bourse, 1930), p. 127.

2. Machine Aesthetics

1. See the report of the U.S. Commission, *International Exposition of Modern Decorative and Industrial Art in Paris, 1925* (Washington, D.C.: Government Printing Office, 1925), p. 18.

2. Joseph Sinel first used the term on his business cards, according to Sheldon Cheney and Martha Candler Cheney, *Art and the Machine: An Account of Industrial Design in 20th-Century America* (New York: Whittlesey House, 1936), p. 55.

3. Charles R. Richards, *Art in Industry* (New York: Macmillan, 1922), unpaged Preface; Matlack Price, "Talented Craftsmen at Architectural League Exhibit," *Arts and Decoration* 20 (March 1924), 42.

4. Richards described these organizations, pp. 464–67. Price described the 1924 Architectural League exhibit, p. 42.

5. C. Adolph Glassgold, "The Modern Note in Decorative Arts," *The Arts* 13 (April 1928), 224.

6. R. L. Duffus, *The American Renaissance* (New York: Alfred A. Knopf, 1928), pp. 113, 115, 116.

7. See Frank Kingdon, *John Cotton Dana: A Life* (Newark: The Public Library and Museum, 1940).

8. A motto quoted in ibid., p. 115.

9. *The Reminiscences of Holger Cahill* (1957), p. 167, in The Oral History Collection of Columbia University; William Friedman, ed., *20th Century Design: U.S.A.* (Buffalo: Albright Art Gallery, 1959), p. 8.

10. Calvin Tomkins ignores Bach and the industrial art department in his *Merchants and Masterpieces: The Story of the Metropolitan Museum of Art* (New York: E. P. Dutton, 1970). The museum's charter is quoted by Rudolph Rosenthal and Helena L. Ratzka, *The Story of Modern Applied Art* (New York: Harper, 1948), p. 177. On Bach's life see "Richard F. Bach, Arts Adviser, 80," *New York Times*, Feb. 18, 1968, p. 80; Bach, *Museum Service to the Art Industry: An Historical Statement to 1927* (New York: The Metropolitan Museum of Art, 1927), p. 5.

11. Bach, *Museums and the Industrial World* (New York: The Metropolitan Museum of Art, 1926), pp. 1–2. He delivered the speech on May 19, 1920.

12. Paragraph synthesized from two articles by Bach: "A Note on Producers of Industrial Art and Their Relation to the Public," *Arts and Decoration* 18 (Dec. 1922), 82, 84, 96; "What Is the Matter with Our Industrial Art?" *Arts and Decoration* 18 (Jan. 1923), 14–15, 46, 49.

13. Léon V. Solon, "The Fostering of American Industrial Art by the Metropolitan Museum," *The Architectural Record* 58 (Sept. 1925), 291–92.

14. Read, "The Exposition in Paris," *International*

Studio 82 (Nov. 1925), 96.

15. Elbert Francis Baldwin, "The Paris Decorative Arts Exposition," *The Outlook*, July 29, 1925, p. 456.

16. W. H. Scheifley and A. E. du Gord, "Paris Exposition of Decorative Arts," *Current History* 23 (Dec. 1925), 360–62.

17. Richardson Wright, "The Modernist Taste," *House and Garden* 48 (Oct. 1925), 110.

18. Read, "Exposition in Paris," pp. 94, 95.

19. Read, "Creative Design in Our Industrial Art," *International Studio* 83 (March 1926), 56.

20. Edward L. Bernays, *Biography of an Idea: Memoirs of Public Relations Counsel Edward L. Bernays* (New York: Simon and Schuster, 1965), pp. 303–304, 309–12.

21. A complete list of delegates with their affiliations appeared in the report of the U.S. Commission, pp. 6–10.

22. "An Exhibition of Contemporary European Industrial Arts," *Bulletin of the Metropolitan Museum of Art* 21 (Jan. 1926), 2.

23. Joseph Breck, "Modern Decorative Arts: Some Recent Purchases," *Bulletin of the Metropolitan Museum of Art* 21 (Feb. 1926), 37.

24. "Window Display," *Fortune* 15 (Jan. 1937), 92.

25. See "Notes," *Architecture* 58 (July 1928), 21.

26. William H. Baldwin, "Modern Art and the Machine Age," *The Independent*, July 9, 1927, pp. 35–39.

27. As quoted by Baldwin, p. 39.

28. C. Adolph Glassgold, "Modern American Industrial Design," *Arts and Decoration* 35 (July 1931), 30.

29. R. L. Leonard and C. A. Glassgold, eds., *Annual of American Design 1931* (New York: Ives Washburn, 1930), pp. 9–10.

30. "The Home of Yesterday To-day and To-morrow," p. 27.

31. "Industrial Design," p. 79.

32. Rosenthal and Ratzka, *The Story of Modern Applied Art*, p. 162.

33. Frankl, *Form and Re-form: A Practical Handbook of Modern Interiors* (New York: Harper, 1930), pp. 13–19. See also Rosenthal and Ratzka, pp. 172–74.

34. Paul Morand, "Paris Letter," *The Dial* 79 (Oct. 1925), 331–32.

35. David Gebhard, "The Moderne in the U.S. 1920–1941," *Architectural Association Quarterly* 2 (July 1970), 5.

36. Le Corbusier, *Towards a New Architecture*, trans. Frederick Etchells (1927; rpt., New York: Frederick A. Praeger, 1963), p. 33. The following paragraph is based on the entire volume. Quoted phrases are from pp. 7, 9, 7,

102, 192, 14, 224, 224–25, 214, 119.

37. As quoted in "Decorative Artists Form Union," *The Architectural Record* 64 (Aug. 1928), 164.

38. Frankl, "Merchandising the Modern Idea in Decoration," *Advertising and Selling*, Dec. 26, 1928, p. 63.

39. Ferriss, *The Metropolis of Tomorrow* (New York: Ives Washburn, 1929), p. 17.

40. Mumford, "Art in the Machine Age," *The Saturday Review of Literature*, Sept. 8, 1928, p. 102.

41. Kiesler, *Contemporary Art Applied to the Store and Its Display* (New York: Brentano's, 1930), p. 9; Deskey, "The Rise of American Architecture and Design," *The London Studio* 5 (April 1933), 272–73.

42. Dana, "The Cash Value of Art in Industry," *Forbes*, Aug. 1, 1928, p. 17.

43. *Form and Re-form*, p. 31.

44. Ibid., pp. 5, 163. Emphasis mine.

45. *Towards a New Architecture*, p. 10.

46. Ibid., p. 264.

47. Paul T. Frankl, *New Dimensions: The Decorative Arts of Today in Words and Pictures* (New York: Payson and Clarke, 1928), pp. 61, 18–19.

48. "The Moderne in the U.S.," p. 7.

49. *The Metropolis of Tomorrow*, p. 16.

50. Ibid., p. 124.

51. *Ultimo: An Imaginative Narration of Life under the Earth* (New York: E. P. Dutton, 1930), n. p.

52. Ferriss, *Metropolis of Tomorrow*, p. 53. See also Dan Klein, "The Chrysler Building," *The Connoisseur* 185 (April 1974), 294–301.

53. Sheldon Cheney, *The New World Architecture* (New York: Longmans, Green, 1930). Quotations are from pp. 73–74, 76, 377, 265, 97.

54. See "Modern Rooms on Display," *New York Times*, June 19, 1929, p. 64. Walter Rendell Storey's review of the exposition, "Making Modern Rooms 'All of a Piece,'" *New York Times Magazine*, July 7, 1929, pp. 16–17, referred only in passing to Mendelsohn.

55. As quoted in "Frank Lloyd Wright and Hugh Ferriss Discuss This Modern Architecture," *The Architectural Forum* 53 (Nov. 1930), 536.

56. Huxley, "Puritanism in Art," *Creative Art* 6 (March 1930), 200–201.

57. *Eric Mendelsohn: Letters of an Architect*, ed. Oskar Beyer, trans. Geoffrey Strachan (New York: Abelard-Schuman, 1967), p. 32.

58. See Susan King, *The Drawings of Eric Mendelsohn* (Berkeley: University Art Museum, University of California, 1969). On German use of reinforced concrete in the field see

Keith Mallory and Arvid Ottar, *The Architecture of War* (New York: Pantheon, 1973), particularly p. 48.

59. *Mendelsohn: Letters*, p. 36.

60. From a 1926 letter as quoted by King, *Drawings*, p. 47.

61. From an unpublished lecture, "The Laws of Modern Architecture," delivered in 1924, as quoted by King, *Drawings*, p. 59.

62. *Contemporary Art Applied to the Store and Its Display*, pp. 67–68.

63. Bach, "Machanalia," *The American Magazine of Art* 22 (Feb. 1931), 102.

3. The New Industrial Designers

1. On Guild see E. F. Lougee, "From Old to New with Lurelle Guild," *Modern Plastics* 12 (March 1935), 14; Francis Sill Wickware, "Durable Goods Go to Town!" *Forbes*, Nov. 15, 1936, p. 36; "Designer for Mass Production," *Art and Industry* 24 (June 1938), 228; and "Design in New York," *Industrial Design* 7 (Oct. 1960), 69.

2. Don Wallace, *Shaping America's Products* (New York: Reinhold, 1956), p. 66; Sheldon Cheney and Martha Candler Cheney, *Art and the Machine: An Account of Industrial Design in 20th-Century America* (New York: Whittlesey House, 1936), pp. 87–88.

3. Cheney, pp. 77–80; Walter Rendell Storey, "Unification in Modern Design: George Sakier—Interior Decorator," *The Studio* 123 (March 1942), 66–69.

4. "Design in New York," p. 83; J. Roger Guilfoyle, "A Half-Century of Design," *Industrial Design* 18 (June 1971), 45–49.

5. Dorothy Wagner Puccinelli, "Joseph Sinel Industrial Designer," *California Arts and Architecture* 58 (June 1941), 23.

6. "Pioneering in Design," *Art and Industry* 24 (May 1938), 174–78.

7. Linton Wilson, "Gustav Jensen," *Pencil Points* 18 (March 1937), 131–50.

8. Jerry Streichler, "The Consultant Industrial Designer in American Industry from 1927 to 1960" (Ph.D. diss., New York University, 1962), pp. 56–57; "Egmont Arens, 78, Designer, Is Dead," *New York Times*, Oct. 2, 1966, p. 87.

9. Dorothy Grafly, "Peter Muller-Munk: Industrial Designer," *Design* 47 (May 1946), 9; "Industrial Designer Is Ruled a Suicide," *New York Times*, March 14 ,1967, p. 35.

10. E. F. Lougee, "Planning Ahead with Gilbert Rohde," *Modern Plastics* 12 (July 1935), 13–15, 56–58; "Gilbert Rohde," *The National Cyclopaedia of American Biography* 33 (New York: James T. White, 1947), 224; and *A Modern Consciousness: D. J. De Pree / Florence Knoll* (Washington, D.C.: Smithsonian Institution, 1975), pp. 6–8, 12.

11. "Designers of To-day: Russel Wright," *The London Studio* 9 (June 1935), 317–23; "Russel Wright," *Current Biography* (New York: H. W. Wilson, 1940), p. 887; Streichler, "Consultant Industrial Designer," pp. 54–56; and "Biographical Sketch of Russel Wright," prepared by Virginia Burdick Public Relations, c. 1951, Series III, Box 1, Russel Wright Collection, George Arents Research Library for Special Collections at Syracuse University.

12. *Art and the Machine*, pp. 69–76.

13. See Mary Siff, "A Realist in Industrial Design," *Arts and Decoration* 41 (Oct. 1934), 47; Teague, "What Industrial Designers Can Do," *Barron's*, March 31, 1941, p. 20; Teague, "A Quarter Century of Industrial Design in the United States," *Art and Industry* 51 (Nov. 1951), 154–56; and Teague, "The Growth and Scope of Industrial Design in the United States," *Journal of the Royal Society of Arts* 107 (Aug. 1959), 646.

14. As quoted by Earnest Elmo Calkins in *"And Hearing Not—": Annals of an Adman* (New York: Scribner, 1946), p. 207. For biographical details see "Walter Dorwin Teague: Dean of Design," *Printers' Ink*, Jan. 30, 1959, pp. 84–85; "Walter D. Teague, Ad Artist, Industrial Design Leader, Dies," *Advertising Age*, Dec. 12, 1960, p. 68; and "Walter Dorwin Teague," *The National Cyclopaedia of American Biography* 50 (New York: James T. White, 1968), 21–22. His clippings of Parrish and Pyle are in Box WDT Sr. 16, Walter Dorwin Teague Collection, George Arents Research Library for Special Collections at Syracuse University (hereafter referred to as *Teague*).

15. From loose pages folded in a notebook. uncatalogued, *Teague*.

16. Siff, "A Realist in Industrial Design," pp. 44, 46–47.

17. Reproduced in Earnest Elmo Calkins, *The Business of Advertising* (New York: D. Appleton, 1915), p. 206. For other "Teague borders" see "Some Recent Examples of Decorative Design Done by Walter Dorwin Teague," *The Printing Art* 29 (March 1917), n. p.

18. Jay Hambidge, *The Elements of Dynamic Symmetry* (New York: Brentano's, 1926), introduction.

19. Teague, *Design This Day: The Technique of Order in the Machine Age* (New York: Har-

court, Brace, 1940), pp. 145, 159, 127.

20. Uncatalogued manuscript, *Teague*.

21. As paraphrased by interviewer Kenneth Reid, "Walter Dorwin Teague, Master of Design," *Pencil Points* 18 (Sept. 1937), 543.

22. According to a letter from his son, W. Dorwin Teague, to the author, Nov. 3, 1976.

23. Reid, "Walter Dorwin Teague," p. 543. For an illustration see Siff, "A Realist," p. 48.

24. Boxes WDT Sr. 2, 4, 6, 8, 10, 11, 13, 14, 16, *Teague*.

25. Cassette letter from Loewy to the author, Feb. 29, 1976.

26. Siff, "A Realist," p. 46.

27. Teague to Mrs. M. Worth, Oct. 17, 1940, reel 16–10, *Teague*.

28. "Both Fish and Fowl," *Fortune* 9 (Feb. 1934), 90.

29. Biographical facts are mostly from his unfinished autobiography, published in part as *Miracle in the Evening* (Garden City, N.Y.: Doubleday, 1960). Two manuscript versions, which extend beyond *Miracle*'s abrupt 1925 stopping point, are in the Norman Bel Geddes Collection, Hoblitzelle Theatre Arts Library, Humanities Research Center, University of Texas at Austin (hereafter referred to as *Geddes*). Although he was (and is) often referred to as "Bel Geddes" (Bel taken from the name of his first wife, Helen Belle Sneider, to make a collaborative pen name), he preferred simply "Geddes" (pronounced *ge-diss*).

30. From a synthesis read by Geddes, Hiram Kelly Moderwell's *Theatre of Today* (1914; rpt., New York: Dodd, Mead, 1927), p. 124. On European theater in the early twentieth century and the introduction of its ideas to America see Oscar G. Brockett, *History of the Theatre* (Boston: Allyn and Bacon, 1968), especially pp. 366–73, 591–99, 630–35.

31. Model photographs appeared in Geddes, *A Project for a Theatrical Presentation of the Divine Comedy of Dante Alighieri* (New York: Theatre Arts, 1924).

32. The scored passage is on p. 3 of Geddes's copy of *Structures and Sketches*, trans. Herman George Scheffauer (Berlin: Ernst Wasmuth, n.d.). Both the book and the sketch, pasted inside the front cover by Geddes, were dated Nov. 25, 1924, by Mendelsohn.

33. From a tear sheet, Sources 71—APq.–14., *Geddes*.

34. See documents in file 314, *Geddes*, regarding his support by the fair's design board, including most of New York's modernistic architects, and by Frank Lloyd Wright, as well as his unsuccessful attempt to become licensed in New York. Robert A. M. Stern has argued

that industrial design seriously threatened the architectural profession in this period. See his "Relevance of the Decade," *Journal of the Society of Architectural Historians* 24 (March 1965), 7.

35. Design Course/SC–6/y.–2., *Geddes*.

36. From MS "Starting of the Industrial Design Profession," July 7, 1953, autobiography, Jamaica version, chs. 69–70, *Geddes*. See file 134.

37. Geddes to Waite, June 29 [1928], in a file labeled "Personal Correspondence between Norman Bel Geddes and Frances," *Geddes*.

38. Geddes, "Designing the Office of Today," *Advertising Arts*, Jan. 1931, pp. 47–51. See also file 133, *Geddes*; Walter Rendell Storey, "The Decorator as a Minister of Trade," *New York Times Magazine*, June 23, 1929, pp. 16, 19; "A Modern Room for a Modern Purpose," *The American Architect*, July 20, 1929, pp. 96–102; and "A Dramatic Background for Modern Business," *Theatre Arts Monthly* 13 (Sept. 1929), 712 and facing page.

39. File 136, *Geddes*. In his *Horizons* (Boston: Little, Brown, 1932), pp. 233–34, Geddes stressed the honesty of his use of materials in this project even though he himself provided designs for Simmons's return to imitation-wood finishes before 1932.

40. Files 152, 153, *Geddes*.

41. Paraphrased report of phone conversation with Toledo Scale president Hugh D. Bennett, Feb. 29, 1929, file 153, *Geddes*.

42. *Horizons*, p. 206.

43. Some years later the company constructed a new factory from plans provided by industrial architect Albert Kahn. See "Design for Mass Production," *Architectural Record* 87 (Feb. 1940), 86–90.

44. Minutes of meeting at Graham-Paige Company, May 21, 1929, file 161, *Geddes*.

45. As quoted by Munro Innes of *Automobile Topics* in a manuscript written in collaboration with the Geddes office and intended as publicity, file 161, *Geddes*.

46. According to Innes.

47. In *Horizons* (p. 54) Geddes described the design as the least radical of the series and designated it Motor Car Number 1 to distinguish it from his later streamlined teardrop designs. But copy written in 1929 to accompany a photograph of a model of the car clearly identified it as the most radical of the series (file 161, *Geddes*).

48. Geddes, "The Artist in Industry: An Unrepentant Confession by a Famous Scene Designer," *Theatre Guild Magazine* 7 (Jan. 1930), 23–24, 61.

49. Undated notes in folder, "*Miracle in the Evening*: Correspondence of NBG with Publisher," *Geddes*.

50. "Questions and Answers to PM Interview (Selma Robinson)," Jan. 22, 1942, autobiography, ch. 35, *Geddes*.

51. Fuller, *Ideas and Integrities: A Spontaneous Autobiographical Disclosure*, ed. Robert W. Marks (1963; rpt., New York: Collier Books, 1969), pp. 76–78; Fuller's Foreword to Henry Dreyfuss, *Designing for People*, rev. ed. (1967; rpt., New York: Viking Compass, 1974). The name is pronounced *dry-fuss*.

52. Dreyfuss, "What Is Happening in Design," *Electrical Manufacturing* 55 (Feb. 1936), 19.

53. William Hewitt of John Deere and Co., a contributor to "Henry Dreyfuss 1904–1972," *Industrial Design* 20 (March 1973), 37.

54. Letter from Tomajan to ———, Nov. 9, 1953, 1973.15.5, Henry Dreyfuss Symbols Archive, Cooper-Hewitt Museum of Design, Smithsonian Institution, New York (hereafter referred to as *Dreyfuss*).

55. Noted by Beverly Smith, "He's into Everything," *The American Magazine* 113 (April 1932), 43.

56. His associate Rita Hart recalled in a letter to the author, Nov. 4, 1976, that Dreyfuss consciously maintained this image after a reporter noted the effect of his penchant for brown suits.

57. Nelson, in "Henry Dreyfuss 1904–1972," pp. 42–43.

58. *Designing for People*, p. 31.

59. Ibid., pp. 21–22. The date of composition of this credo is uncertain.

60. Transcript, interview conducted by Selma Robinson of *PM*, Jan. 27, 1942, autobiography, Jamaica version, ch. 76, *Geddes*.

61. Unless otherwise noted, biographical information is from Beverly Smith, "He's into Everything," pp. 43, 150–52; Carlton Atherton, "Henry Dreyfuss . . . Designer," *Design* 36 (Jan. 1935), 4–7; J. D. Ratcliff, "Designer for Streamlined Living," *Coronet* 22 (June 1947), 22–26; Alva Johnston, "Nothing Looks Right to Dreyfuss," *The Saturday Evening Post*, Nov. 22, 1947, pp. 20–21, 132, 134–35, 137, 139; "Henry Dreyfuss," *Current Biography* (New York: H. W. Wilson, 1948), pp. 159–61; "Brown Book," 1972.88.178, *Dreyfuss*; letter from Ann Dreyfuss to the author, Jan. 24, 1977; and a cassette letter from Ann Dreyfuss to the author, March 8, 1977.

62. Adler, *The Moral Instruction of Children* (1892; rpt., New York and London: D. Appleton, 1927), pp. 218, 186.

63. As quoted by Samuel Frederick Bacon, *An Evaluation of the Philosophy and Pedagogy of Ethical Culture* (Washington, D.C.: Catholic University of America, 1933), p. 51.

64. Henry Neumann, *Education for Moral Growth* (New York and London: D. Appleton, 1923), p. 263.

65. Ibid., pp. 198–99; Bacon, *Evaluation*, pp. 66–67; Johnston, "Nothing Looks Right," p. 21; cassette letter from Ann Dreyfuss to the author, March 8, 1977.

66. Dreyfuss, *Designing for People*, p. 16.

67. Brecker to Rita Hart, Nov. 12, 1953, 1973.15.5, *Dreyfuss*.

68. Quotations are from discussion by William Fondiller in E. Grosvenor Plowman, *Fashion, Style and Art Spread to Other Lines of Business*, General Management Series, no. 106 (New York: American Management Association, 1929), pp. 17–18. See also Wilson, "Gustav Jensen," p. 135; Wallance, *Shaping America's Products*, pp. 34–40; and *Designing for People*, pp. 92–93.

69. Smith, "He's into Everything," p. 152.

70. Atherton, "Henry Dreyfuss . . . Designer," p. 7.

71. Undated note, 1973.15.34, *Dreyfuss*.

72. Nov. 14, 1931, 1973.15.32, *Dreyfuss*.

73. Bill Davidson, "You Buy Their Dreams," *Collier's*, Aug. 2, 1947, p. 23.

74. Loewy, *Never Leave Well Enough Alone* (New York: Simon and Schuster, 1951), p. 4

75. Autobiography, Feb. 20, 1953, ch. 60, *Geddes*.

76. As quoted by J. Roger Guilfoyle, "A Thousand in One," *Industrial Design* 17 (June 1970), 33.

77. Claude Yelnick, "Raymond Loewy, magicien du progrès," *France Illustration*, no. 402, Sept. 1953, pp. 25–31, 33; Lois Frieman Brand, *The Designs of Raymond Loewy* (Washington: Smithsonian Institution Press, 1975), pp. 47–48; John Kobler, "The Great Packager," *Life*, May 2, 1949, p. 116; "Raymond (Fernand) Loewy," in *Current Biography* (New York: H. W. Wilson, 1941), p. 523; and Sarah Booth Conroy, "If It's Sleek and Floats, Chugs, Drives or Orbits (Among Other Things) Think Raymond Loewy," *Washington Post*, Feb. 23, 1975, section F, p. 3.

78. *Never Leave Well Enough Alone*, p. 10. Loewy emphasized the personal nature of his crusade against vulgarity in a talk at the Renwick Gallery, Smithsonian Institution, Aug. 1, 1975.

79. Quotations are from a cassette letter to the author, Feb. 29, 1976.

80. Unless otherwise noted, biographical information comes from *Never Leave Well Enough Alone*, which is careless with dates. Loewy's birth date is from *Current Biography*, p. 522. For a discussion of his education, see Jeffrey L.

Meikle, "Technological Visions of American Industrial Designers, 1925–1939" (Ph.D. diss., University of Texas at Austin, 1977), pp. 310–12, 844–46.

81. Renwick Gallery talk.

82. Kobler, "The Great Packager," p. 120.

83. Ibid., p. 122.

84. Renwick Gallery talk.

85. Letter from Loewy to the author, April 15, 1976.

86. *Never Leave Well Enough Alone*, p. 83.

87. On the Gestetner job see E. F. Lougee, "Raymond Loewy Tells Why," *Modern Plastics* 12 (Jan. 1935), 21–22; *Never Leave Well Enough Alone*, pp. 81–84; and "A Duplicator, a Designer, and a 28th Anniversary," *Industrial Design* 7 (Feb. 1960), 54–55.

88. Renwick Gallery talk; cassette letter to the author, Feb. 29, 1976.

89. See "Another Way of Obtaining 'Appeal' in the Product," *Product Engineering* 2 (Feb. 1931), 85; *Official Gazette of the United States Patent Office*, June 30, 1931, p. 1160; and *Never Leave Well Enough Alone*, p. 94. Loewy verified his employment by Westinghouse in a letter to the author, April 15, 1976. R. W. Dodge of Westinghouse reported in a letter to the author, Jan. 18, 1977, being unable to track down details of the association.

90. Storey, "The Decorator as a Minister of Trade," p. 16. See also Edward Alden Jewell, "No News Is Cool News," *New York Times*, June 2, 1929, section 9, p. 16; Sidney Blumenthal, "Art in Manufacture," *The American Magazine of Art* 21 (Aug. 1930), 441.

91. Loewy's upholstery work for Hupp was mentioned by Davidson, "You Buy Their Dreams," p. 23. The date of Loewy's initial employment by Hupp and the extent of his early work for the firm are uncertain. Recently he stated that he began working for Hupp early in 1932, according to a quotation in "Aerodynamic Hupp," *Special-Interest Autos* 3 (April–May 1972), 30. But "No News Is Cool News" in 1929 reported him stating that he had just finished some designs for Hupp. Lougee, "Raymond Loewy Tells Why," p. 54, stated that Loewy designed both exterior and interior of the 1926 Hupmobile, but no other source offers verification.

92. Design patents 78,038, *Official Gazette*, March 19, 1929, p. 550; 79,147, *Official Gazette*, Aug. 6, 1929, p. 59; 79,974, *Official Gazette*, Nov. 26, 1929, p. 840; and 80,844, *Official Gazette*, April 1, 1930, p. 57.

93. See Cheney, *Art and the Machine*, p. 106; *Never Leave Well Enough Alone*, pp. 84–89;

and Brand, *The Designs of Raymond Loewy*, pp. 15, 24. Loewy filed for a design patent covering the future 1931 prototype on Aug. 28, 1930. See *Official Gazette*, Feb. 3, 1931, p. 58.

4. Selling Industrial Design

1. Nathan George Horwitt, "Plans for Tomorrow: A Seminar in Creative Design," *Advertising Arts*, July 1934, p. 29. Rita Hart used the phrase in a letter to the author, Nov. 4, 1976.

2. Stanley Nowak, "The Dollars and Cents Angle of Product Design," *Printers' Ink Monthly*, 26 (Feb. 1933), 68.

3. Norman Bel Geddes, *Horizons* (Boston: Little, Brown, 1932), p. 3.

4. Broadus Mitchell, *Depression Decade: From New Era Through New Deal 1929–1941* (New York: Rinehart, 1947), pp. 259–61.

5. Information on numbers of clients is from "Designs of Walter Dorwin Teague," a list provided by his son W. Dorwin Teague; the "Brown Book," *Dreyfuss*; and Geddes, "Affidavit Regarding Activities," July 1946, *Geddes*.

6. Dorfman, *The Economic Mind in American Civilization* 5 (New York: Viking, 1959), 630–31.

7. *Fortune* 6 (Dec. 1932), 94.

8. Maurice Levin, Harold G. Moulton, and Clark Warburton, *America's Capacity to Consume* (Washington: Brookings Institution, 1934), p. 126. Conclusions based on it and its predecessor, *America's Capacity to Produce*, are on pp. 125–33.

9. On Technocracy see Henry Elsner, Jr., *The Technocrats: Prophets of Automation* (Syracuse, N.Y.: Syracuse University Press, 1967); William E. Akin, *Technocracy and the American Dream: The Technocrat Movement, 1900–1941* (Berkeley: University of California Press, 1977).

10. "Preface," *A Philosophy of Production: A Symposium* (New York: The Business Bourse, 1930), pp. viii–ix.

11. Young, "Humanizing Modern Production," pp. 3–28.

12. Baruch, "A Plan for the Regulation of Production," pp. 93, 95.

13. Ford with Samuel Crowther, "The American Way to Wealth and Happiness," pp. 31–44.

14. Abbott, "Obsolescence and the Passing of High-Pressure Salesmanship," pp. 169, 173.

15. Frederick, "Whither Production?" p. 259.

16. Calkins, "The New Consumption Engineer,

and the Artist," pp. 107–29. Quotations from pp. 113, 111, 109.

17. Ibid., pp. 117, 120, 121–22, 124, 115, 125–26.

18. Calkins, "What Consumer Engineering Really Is," in Sheldon and Arens, *Consumer Engineering* (New York: Harper, 1932), p. 1.

19. *Consumer Engineering*, p. 210.

20. On factory management see Samuel Haber, *Efficiency and Uplift: Scientific Management in the Progressive Era 1890–1920* (Chicago: University of Chicago Press, 1964); Edwin T. Layton, Jr., *The Revolt of the Engineers: Social Responsibility and the American Engineering Profession* (Cleveland: Case Western Reserve University Press, 1971), pp. 134–53.

21. See John Chynoweth Burnham, "The New Psychology: From Narcissism to Social Control," in *Change and Continuity in Twentieth-Century America: The 1920's*, ed. John Braeman, Robert H. Bremner, and David Brody (Columbus: Ohio State University Press, 1968), p. 390.

22. *Consumer Engineering*, p. 210.

23. Ibid., p. 95.

24. Resor as quoted by Lucille Terese Birnbaum, "Behaviorism: John Broadus Watson and American Social Thought, 1913–1933" (Ph.D. diss., University of California at Berkeley, 1965), p. 113.

25. Watson, Foreword to Henry C. Link, *The New Psychology of Selling and Advertising* (New York: Macmillan, 1934), p. vii.

26. *Consumer Engineering*, p. 55.

27. Ibid., p. 56.

28. Bush, *The Streamlined Decade* (New York: George Braziller, 1975), p. 22.

29. See *Consumer Engineering*, especially pp. 52–53, 133–34.

30. Ibid., pp. 51, 56–58, 194.

31. *Horizons*, p. 45.

32. William H. Lough with Martin R. Gainsbrugh, *High-Level Consumption: Its Behavior; Its Consequences* (New York: McGraw-Hill, 1935), p. 115.

33. Sheldon Cheney and Martha Candler Cheney, *Art and the Machine: An Account of Industrial Design in 20th-Century America* (New York: Whittlesey House, 1936), pp. 138–39.

34. Link, *The New Psychology of Selling and Advertising*, p. xiii.

35. William J. Acker, "Design for Business," *Design* 40 (Nov. 1938), 12.

36. Worthen Paxton, as paraphrased by Walter McQuade in "An Industrial Designer with a Conspicuous Conscience," *Fortune* 68 (Aug. 1963), 136.

37. Calkins, "Design and Economic Recovery,"

Advertising Arts, July 1933, p. 11.

38. Donald Wilhelm, "The Millstone of Style," *New Outlook* 163 (April 1934), 38.

39. "Standard Practice: Service Department," file 940, *Geddes*.

40. Abbott Kimball, "Building Goodwill with Good Design," *Advertising Arts*, November 1931, pp. 13–14.

41. Flannery, "Styling Modern Merchandise," *Advertising and Selling*, April 16, 1930, p. 68.

42. Unidentified executive quoted by Geoffrey Holme, *Industrial Design and the Future* (London: The Studio, 1934), p. 33. Of six American respondents to a questionnaire sent out by Holme, three were admen.

43. From an advertisement in *Advertising Arts*, April 2, 1930, p. 12.

44. Kimball, "A Design Engineer—Every Business Needs One Today," *Printers' Ink*, April 16, 1931, p. 105.

45. See "The Molded Clock Case Goes Modern with Electric Mechanism," *Plastics and Molded Products* 6 (Aug. 1930), 471; C. W. Luhr and K. G. Berggren, "Originality and Refinement in Electrical Appliances," *Product Engineering* 1 (Oct. 1930), 474.

46. Sinel, "Design Impels Consumer Response," *The Bulletin of the American Ceramic Society* 13 (Nov. 1934), 290.

47. Horwitt, "Plans for Tomorrow," p. 29.

48. "The Eyes Have It," *The Business Week*, Jan. 29, 1930, p. 30; Franklin S. Clark, "Modern Designs Sell," *Forbes*, Aug. 15, 1930, p. 20.

49. "Bel Geddes," *Fortune* 2 (July 1930), 51, 54, 57.

50. "Both Fish and Fowl," *Fortune* 9 (Feb. 1934), 40, 43, 88.

51. Reported in a letter from John S. Tomajan to ———, Nov. 9, 1953, 1973.15.5, *Dreyfuss*.

52. Quoted and cited by Pauline Arnold, "New Designs in Old Bottles," *Printers' Ink Monthly* 21 (Sept. 1930), 44.

53. "Improved Design Would Help Coin Machines," Nov. 1933, pp. 5, 57.

54. See Blood, "Redesign: What It Means in the Comeback Process," *Printers' Ink*, March 30, 1933, pp. 69, 72–74; Blood, "100,793 Women 'Bossed' Our 1933 Product Design Job," *Sales Management*, Sept. 15, 1933, pp. 261, 293.

55. "Maine Mfg. Co. Starts Comeback with Complete Redesign Program," *Sales Management*, Dec. 15, 1936, p. 1040.

56. Maloney, "Case Histories," *Advertising Arts*, July 1934, pp. 34–35; Maloney, "Case Histories in Product Design," *Product Engineering* 5 (March 1934), 96–97; Dreyfuss, "The Profile of Industrial Design," *Machine Design*,

June 22, 1967, p. 158.

57. Donald R. Dohner, in *Product Engineering* 1 (Sept. 1930), 427–29.

58. Otto Kuhler, "Needed in the Product—Art Expression," *Product Engineering* 2 (May 1931), 214.

59. Luhr and Berggren, "Originality and Refinement in Electrical Appliances," pp. 472, 474.

60. "Unemployment and Product Engineers," *Product Engineering* 1 (Dec. 1930), 547. See also Condit, "No Birds in Last Year's Nest," Nov. 1930, p. 530.

61. Condit, "Appearance Counts," *Product Engineering* 2 (Sept. 1931), 418.

62. Brady, "Product Design for Increased Utility and Improved Marketability," *Mechanical Engineering* 53 (Sept. 1931), 675–76.

63. Walter Dorwin Teague (as told to Charles G. Muller), "Modern Design Needs Modern Merchandising," *Forbes*, Feb. 1, 1932, p. 15.

64. Teague, "The Artist in Industry: What He Does and How He Works," *Product Engineering* 3 (June 1932), 245–46.

65. Brady, "Appearance as a Sales Factor in Design," *The Iron Age*, Oct. 27, 1932, p. 648, supp. 22.

66. *Product Engineering* 3 (Dec. 1932), 473.

67. Sinel, "Artist Joins Engineer in the Design of the Industrial Product," *The Iron Age*, Dec. 28, 1933, p. 25.

68. Kuhler, "What Price Beauty of Line and Color," *Metal Progress* 26 (Aug. 1934), 23–25. Emphasis mine. Quoted phrases in the following paragraph are from p. 25.

69. Hervey L. MacCowan, "A Straw in the Wind," *Modern Plastics* 12 (April 1935), 65.

70. Raymond B. Stringfield, *Plastics and Molded Products* 7 (Oct. 1931), 547.

71. Franklin E. Brill, "Some Hints on Molded Design," *Plastic Products* 9 (April 1933), 54.

72. "The Molded Clock Case Goes Modern with Electric Mechanism," p. 471.

73. *Plastics and Molded Products* 6 (Dec. 1930), 696.

74. *Plastics and Molded Products* 6 (Sept. 1930), 532.

75. J. Harry Dubois, *Plastics History U.S.A.* (Boston: Cahners, 1972), p. 185.

76. *Plastics and Molded Products* 8 (Sept. 1932), cover.

77. "The Engineer Meets the Customer," *Product Engineering* 2 (Feb. 1931), 49. Skeptic George S. Brady considered lower production cost the only valid argument for redesign: "Designing the Product for Tomorrow's Industrial Market," *The Iron Age*, Sept. 8, 1932, pp. 363–64.

78. Based on a table provided by Vergil D. Reed, "United States Statistics Record Growth," *Modern Plastics* 16 (Oct. 1938), 11.

79. Brill, "Some Hints on Molded Design," *Plastic Products* 9 (April 1933), 54–55; Brill, "Design for Black," *Modern Plastics* 12 (March 1935), 16–17, 55, 57–58.

80. Harold Van Doren, "A Designer Speaks His Mind," *Modern Plastics* 12 (Sept. 1934), 24.

81. The Loewy ad appeared in *Plastic Products* 10 (June 1934), 205. Other ads in the campaign appeared in *Plastic Products* 10 (Jan. 1934), 5; (Feb. 1934), 47; (March 1934), 88; (April 1934), 124; (May 1934), 164; 11 (July 1934), 6; (Aug. 1934), 43; and *Modern Plastics* 11 (Oct. 1934), 5. The series ran in *Sales Management* in 1933–1934.

82. Typical advertisements in *Plastic Products* 10 (June 1934), 203; (June 1934), between 220–21; and 11 (July 1934), 8.

83. "Industrial Design Comes of Age," *Business Week*, May 23, 1936, p. 17.

84. See "Current Activity in Product Development," *Product Engineering* 3 (Nov. 1932), 436–37; "The 1935 Survey of Changes and Trends in Product Design," 6 (Nov. 1935), 426.

85. As quoted by Seymour Freedgood, "Odd Business, This Industrial Design," *Fortune* 59 (Feb. 1959), 132.

86. "Industrial Design Comes of Age," p. 17.

87. Van Doren, "Industrial Design and the Manufacturer," *Design* 38 (June 1936), 20.

88. *Horizons*, p. 242.

89. *Modern Plastics* 15 (Oct. 1937), 122–23.

90. Vassos, "'Have Designers Missed a Bet?'" *Creative Design* 3 (Oct.–Nov. 1938), 18.

91. Lougee, "What Is Industrial Design," *Modern Plastics* 15 (Sept. 1937), 26.

5. Industrialized Design

1. Loewy, "Modern Metals in Modern Designing," *The Iron Age*, June 13, 1935, p. 29. I relied as well on a publicity release, "Corner of an Industrial Designer's Office and Studio," from the files of Raymond Loewy International Inc., New York City (hereafter referred to as *Loewy*).

2. Letter from Bach to Geddes, Oct. 27, 1932, autobiography, Jamaica version, ch. 68, *Geddes*.

3. Contract, Sept. 30, 1930, file 199, *Geddes*.

4. "Standard Practice of Norman Bel Geddes and Company," 1945, file 940, *Geddes*.

5. Ibid.

6. "Standard Practice," Production Department, 1945, file 940, *Geddes*.

7. "Standard Practice of Norman Bel Geddes and Company," 1945, file 940, *Geddes*.

8. "Standard Practice," Production Department, 1945, file 940, *Geddes*; Kenneth Reid, "Masters of Design: 2—Norman Bel Geddes," *Pencil Points* 18 (Jan. 1937), 13.

9. Letter from Garth Huxtable to the author, Nov. 5, 1976. Geddes's printed checklists of items to be included in luggage on business trips were complete to the number of paper clips.

10. This paragraph is based on "Both Fish and Fowl," *Fortune* 9 (Feb. 1934), 43, 94; Sheldon Cheney and Martha Candler Cheney, *Art and the Machine: An Account of Industrial Design in 20th-Century America* (New York: Whittlesey House, 1936), pp. 62–69; Reid, "Masters of Design," pp. 13, 17; a one-page biography of Worthen Paxton, 1945, autobiography, Jamaica version, ch. 69, 70, *Geddes*; Roger L. Nowland, "Planning Products with Facts," *Distribution Age* 45 (June 1946), 32; Irwin Ross, *The Image Merchants: The Fabulous World of Public Relations* (Garden City, N.Y.: Doubleday, 1959), pp. 90–91; "Design in New York," *Industrial Design* 7 (Oct. 1960), 76, 105; Jerry Streichler, "The Consultant Industrial Designer in American Industry from 1927 to 1960" (Ph.D. diss., New York University, 1962), pp. 98–99; and letters to the author from Garth Huxtable, Nov. 5, 1976, and Carl Otto, Dec. 7, 1976, and March 14, 1977.

11. "Both Fish and Fowl," p. 43; letter from Rita Hart to the author, Nov. 4, 1976.

12. Teague to Powel Crosley, Jr., March 17, 1937, reel 16–3, *Teague*.

13. This paragraph is based on "Both Fish and Fowl," p. 43; Kenneth Reid, "Walter Dorwin Teague, Master of Design," *Pencil Points* 18 (Sept. 1937), 544; Eugene A. Hosansky, "Walter Dorwin Teague Associates," *Industrial Design* 13 (June 1966), 54; a letter from W. Dorwin Teague, Jr., to the author, Nov. 3, 1976; and a survey of Teague's files.

14. Loewy, *Never Leave Well Enough Alone* (New York: Simon and Schuster, 1951), p. 130.

15. Cassette letter from Loewy to the author, Feb. 29, 1976. For other details see Cheney, *Art and the Machine*, pp. 88–90; "Designs by Loewy: From Packages to Modern Stores That Sell Them," *Printers' Ink*, May 29, 1959, p. 73; and Charles M. Macko, "Business Consults the Industrial Designer," *Barron's*, July 15, 1940. p. 20.

16. On the Big Ben case, see "Now the Beauty Engineers," *Popular Mechanics Magazine* 58 (Oct. 1932), 555–56; Dreyfuss, *Ten Years of Industrial Design* (New York: privately printed, 1939), n.p.; and a memo from J.[ulian] E.[verett] to Rita Hart, June 1953, 1973.15.5, *Dreyfuss*. The fullest account of the design process appeared in Harold L. Van Doren's *Industrial Design: A Practical Guide* (New York: McGraw-Hill, 1940).

17. C. M. Morley to Teague, March 24, 1931, reel 16–8, *Teague*.

18. C. F. Norton to Teague, Jan. 16, 1936, reel 16–7, *Teague*.

19. Autobiography, Oct. 10, 1955, ch. 79, *Geddes*.

20. *Never Leave Well Enough Alone*, p. 135.

21. According to a Sears executive quoted by Streichler, "The Consultant Industrial Designer," p. 54.

22. Teague to Ford, June 14, 1934; Ford to Teague, Sept. 28, 1934; reel 16–6, *Teague*.

23. "Standard Practice," Service Department, 1944, file 940, *Geddes*.

24. "Notes for Boston Lecture," March 14, 1933, 1973.15.22(a), *Dreyfuss*.

25. Teague to Edward P. Bailey, Jr., April 3, 1931, reel 16–8, *Teague*.

26. Bailey to Teague, April 2, 1931, reel 16–8, *Teague*.

27. "Preliminary Estimate," file 260, *Geddes*.

28. Contract, Sept. 30, 1930, and Martin Dodge, "Job History," Jan. 23, 1946, file 199, *Geddes*.

29. Memo from John D. Brophy to Teague, Nov. 21, 1936, reel 16–6, *Teague*.

30. Geddes to Frances Waite Geddes, Jan. 5, 1939, "Personal Correspondence between Norman Bel Geddes and Frances," *Geddes*.

31. Cassette letter from Loewy to the author, Feb. 29, 1976.

32. *Never Leave Well Enough Alone*, p. 127.

33. Bill Davidson, "You Buy Their Dreams," *Collier's*, Aug. 2, 1947, p. 23.

34. "Standard Practice," Service Department, 1944, file 940, *Geddes*.

35. Advertisement for Horse Head Zinc, *Advertising Arts*, Sept. 1934, between pp. 36–37; "Designing to Sell," *Sales Management*, Oct. 20, 1934, p. 404.

36. C. F. Norton to Teague, Jan. 16, 1936, and Martin Dodge to Norton, Jan. 18, 1936, reel 16–7, *Teague*.

37. "Inviting Sales Through Redesign," *Modern Plastics* 13 (June 1936), 16.

38. T. J. Maloney, "Case Histories," *Advertising Arts*, March 1934, p. 33.

39. See also "Design for Dollars," *Business Week*, Nov. 18, 1933, p. 12; Rex F. Clarke, "Heater Redesign Jumps Sales 400% for American Gas Machine," *Sales Management*, Dec. 1, 1933, pp. 532–33; and Davidson, "You Buy Their Dreams," p. 23.

40. Hanson to Teague, Aug. 1, 1936, reel 18A, *Teague*.

41. Hanson to Teague, Aug. 18, 1936, reel 18A, *Teague*.

42. Hanson to Teague, Feb. 24, 1937, reel 18A, *Teague*.

43. Memo from "J" to Teague, Aug. 1, 1938, reel 18A, *Teague*.

44. Teague, "Why Disguise Your Product?" *Electrical Manufacturing*, Oct. 1938, p. 46.

45. Hanson to John D. Brophy, Dec. 29, 1938, reel 18A, *Teague*.

46. "Notes for Boston Lecture," March 14, 1933, 1973.15.22(a), *Dreyfuss*.

47. Harold Van Doren, "Industrial Design and the Manufacturer," *Design* 38 (June 1936), 20–21.

48. Neil H. Borden, *The Economic Effects of Advertising* (Chicago: Richard D. Irwin, 1942), pp. 122–23.

49. Roberts to Earl Newsom, Dec. 14, 1933, file 267, *Geddes*.

50. Statistics are from Andrew W. Cruse, "The Electrical Goods Industries," in *Technological Trends and National Policy: Including the Social Implications of New Inventions*, comp. U.S. National Resources Committee, Subcommittee on Technology (Washington, D.C.: Government Printing Office, 1937), p. 317; Borden, pp. 397–98; and U.S. Bureau of the Census, *Historical Statistics of the United States: Colonial Times to 1957* (Washington, D.C.: Government Printing Office, 1960), p. 510.

51. *Mechanization Takes Command: A Contribution to Anonymous History* (1948; rpt., New York: W. W. Norton, 1969), p. 568.

52. "The Engineer Meets the Customer," *Product Engineering* 2 (Feb. 1931), 49.

53. Otis Pease, *The Responsibilities of American Advertising: Private Control and Public Influence, 1920–1940* (New Haven: Yale University Press, 1958), p. 101.

54. W. H. MacHale, "Let's Look at Radio Cabinets," *Modern Plastics* 15 (Feb. 1938), 21. MacHale was arguing for plastic cases, but the rationale held for the metal cases of a few years before. See also "1,250,000 Out of 4,200,000 U.S. Radios," *Fortune* 11 (Feb. 1935), 164, 166; "The Engineer Meets the Customer," p. 50.

55. "The Baby Radio," *Fortune* 8 (July 1933),

64–65; W. E. Freeland, "The Importance of the Product in the Recovery Period," *Mechanical Engineering* 56 (May 1934), 278; "1,250,000 Out of 4,200,000 U.S. Radios," p. 168; and "Radio III: A $537,000,000 Set Business," *Fortune* 17 (May 1938), 118.

56. Franklin E. Brill, "Midget Radios Versus Electric Clocks," *Plastic Products* 9 (July 1933), 182.

57. As reported in "Designing Things Better," *Business Week*, Nov. 7, 1936, p. 41.

58. Alvin Kabaker, "Better Design for America," *Design* 37 (Dec. 1935), 36–37; Jane Fiske Mitarachi, "Design in the Chicago Midwest," *Industrial Design* 3 (Oct. 1956), 73–74.

6. Everything from a Match to a City

1. Teague, "What Industrial Designers Can Do," *Barron's*, March 31, 1941, p. 20.

2. Loewy's letter to the author of April 15, 1976, did not specify whether percentages referred to income, time, or number of commissions.

3. The following account is based on "New Design Opens New Outlets," *Printers' Ink*, April 6, 1933, pp. 51–52; "Both Fish and Fowl," *Fortune* 9 (Feb. 1934), 42; and file 267, *Geddes*.

4. W. Frank Roberts to Geddes, Sept. 5, 1933, file 267, *Geddes*.

5. For pamphlets and advertisements see Geddes's scrapbooks for 1932–1934.

6. "Oriole Bel Geddes Continues to Create Interest," *The American Enameler*, May 1934, a clipping in Geddes's 1934 scrapbook.

7. Dreyfuss, "Good Design Pays," *Electrical Manufacturing*, May 1934, p. 55. See also R. S. McFadden, "Designing to Sell," *Sales Management*, May 15, 1933, pp. 512–13, 537; "Best-Dressed Products Sell Best," *Forbes*, April 1, 1934, p. 13.

8. Dreyfuss, "Business Adopts Design (Part II)," *The Enamelist*, Nov. 1933, p. 18.

9. On GE refrigerator styling see Don Wallance, *Shaping America's Products* (New York: Reinhold, 1956), pp. 70–71, 73–75.

10. Transcript of Geddes interview by Selma Robinson of *PM*, Jan. 27, 1942, autobiography, Jamaica version, ch. 76, *Geddes*.

11. From a radio script for NBC's "Rex Cole Mountaineers," March 29, 1934, 1973.15.22(b), *Dreyfuss*. See also Dreyfuss, "Good Design Pays," p. 54.

12. "Keeping a Step Ahead of Competitors: Notes on the Case History of the 'Coldspot' Refrig-

erator Developed by Raymond Loewy," c. 1939, *Loewy*. See also "Up from the Egg," *Time*, Oct. 31, 1949, p. 72.

13. "Keeping a Step Ahead of Competitors," *Loewy*.

14. "Sears 1937 Coldspot Replete with Mechanical and Design Changes," *Loewy*.

15. The phrase is from Carl Otto's letter to the author of March 14, 1977.

16. "Keeping a Step Ahead of Competitors," *Loewy*.

17. Promotional material, *Loewy*.

18. Harold Van Doren, *Industrial Design: A Practical Guide* (New York: McGraw-Hill, 1940), p. 355.

19. James F. Lincoln, "Appearance a Vital Factor in Machine Design: The Influence of the Electric Arc Welding Process on the Design of Machinery," *Machinery* 36 (April 1930), 585, 587.

20. P. Huber, "Styled Machines Sell Well," *Electrical Manufacturing* 18 (Oct. 1936), 70–72.

21. Dreyfuss, "What Is Happening in Design," *Electrical Manufacturing*, Feb. 1936, p. 20.

22. Promotional material, *Loewy*.

23. File 291, *Geddes*.

24. John J. Pelley quoted in "New Haven Plans 50 Streamline Coaches, Air Conditioned and Painted in Gay Colors," *New York Times*, Feb. 19, 1934, p. 17.

25. Teague, "Design for the Railroad," *Advertising Arts*, May 1934, pp. 32–33; unpublished account by Teague, reel 16–8, *Teague*. The next two paragraphs are based on the same sources.

26. Teague's seats and luggage racks weighed two tons less per coach than standard ones. Pullman Bradley engineers themselves had cut the weight of a car by one-fourth in switching from high-carbon steel to stainless steel and aluminum.

27. According to a memo from Martin Dodge to Teague, June 2, 1936, reel 16–8, *Teague*.

28. "Notes on Raymond Loewy's Work for the Panama Railway Steamship Company," *Loewy*.

29. "Notes on Use of Glass in S.S. PANAMA," *Loewy*.

30. "Notes on Raymond Loewy's Work," *Loewy*.

31. Gretta Palmer, "The Artist Turns Big Shot," *Today*, June 22, 1936, clipping in *Geddes*.

32. *Horizons* (Boston: Little, Brown, 1932), pp. 84–109.

33. Van Doren, *Industrial Design*, p. 350.

34. "Yawman and Erbe Designs Desks to Please the President," *Sales Management*, April 9, 1932, p. 40.

35. Teague, "To Improve Economy, Performance and Appearance," *Electrical Manufacturing*, July 1937, pp. 38–41; "A Sales Office," *The Architectural Record* 79 (Jan. 1936), 73–78.

36. Harry Harvey Porter to the National Cash Register Company, Sept. 28, 1938, reel 16–8, *Teague*.

37. "Main Street, U.S.A.," *The Architectural Forum* 70 (Feb. 1939), 85. The magazine defined modernization as the expenditure of more than $1,000 in a single calendar year.

38. As quoted in "Walter Dorwin Teague Designs Glass," *Design* 37 (March 1936), 40.

39. Teague to Powers Pace, Feb. 1, 1936, reel 16–10, *Teague*. Of the renderings catalogued in drawing list DS#61, reel 7A, only four—two clothing stores, a bar, and a dairy store—have survived, frames 1036–1039, reel 3P. Two of these appeared in Kenneth Reid's "Walter Dorwin Teague, Master of Design," *Pencil Points* 18 (Sept. 1937), 555.

40. Illustrated in Teague, "Machine Age Aesthetics," *Advertising Arts*, July 1932, p. 7.

41. See a memo from Martin Dodge to Teague, Sept. 11, 1936, reel 16–10, *Teague*.

42. "Cushmans Sons Bakeries, New York," *American Architect and Architecture* 150 (June 1937), 95.

43. As quoted in "Symbols of Industry Translated in Design," *Modern Plastics* 13 (April 1936), 17.

44. "Socony-Vacuum Exhibit, RCA Building, New York," *The Architectural Forum* 64 (March 1936), 171–74; "Touring Bureau of Modern Design," *The Architect and Engineer* 126 (July 1936), 56–57.

45. Based on drawing list DS#67, reel 7A, *Teague*; "Standardized Service Stations," *Architectural Record* 82 (Sept. 1937), 69–72; "The Rest-Room Business," *Fortune* 18 (Oct. 1938), 28; and Daniel Schwarz, "Art for Industry's Sake," *New York Times Magazine*, April 2, 1939, p. 19.

46. From the long "Report on Service Stations Prepared for the Socony-Vacuum Oil Company, Inc.," 1934, file 322, *Geddes*. The following account and all quotations are taken from this report, which Geddes delivered to the company, assisted by twenty-four explanatory charts, two models, and fifteen blueprints.

47. Palmer, "The Artist Turns Big Shot," n. p.; Sheldon Cheney and Martha Candler Cheney, *Art and the Machine: An Account of Industrial Design in 20th-Century America* (New York: Whittlesey House, 1936), p. 162. For an idea of the vast difference in complexity between product design and urban planning,

consult Robert A. Caro's *The Power Broker: Robert Moses and the Fall of New York* (New York: Vintage, 1975).

48. Drawing list #S-102, reel 7A, *Teague*, provides the only evidence of this project.

49. See file 263, *Geddes*. Quotations are from Geddes to Rockefeller, June 29, 1933, file 963, and Rockefeller to Geddes, April 3, 1934, file 263. Apparently the project lapsed. See Peter Collier and David Horowitz, *The Rockefellers: An American Dynasty* (New York: Holt, Rinehart and Winston, 1976), p. 174.

50. Industrial designers hardly pioneered the idea of prefabricated housing. For an account of advocates and their systems during the thirties see Alfred Bruce and Harold Sandbank, *A History of Prefabrication* (New York: John B. Pierce Foundation, 1944).

51. Teague to F. Spencer of Eastman Kodak, Feb. 24, 1936, reel 16–4, *Teague*.

52. Le Corbusier to Teague, March 8, 1937, reel 16–3, *Teague*.

53. Teague to F. Stuart Fitzpatrick, Civic Development Department, U.S. Chamber of Commerce, April 29, 1940, reel 16–3, *Teague*.

54. File 400, *Geddes*.

7. The Practical Ultimate

1. Giedion, *Mechanization Takes Command: A Contribution to Anonymous History* (1948; rpt., New York: W. W. Norton, 1969), p. 610.

2. Van Doren, *Industrial Design* (New York: McGraw-Hill, 1940), pp. 45–46, 54.

3. Loewy's letter to the *Times* (London) was reprinted as "Aesthetics in Industry," *The Architectural Review* 99 (Jan. 1946), lv.

4. Sinel, "What Is the Future of Industrial Design?" *Advertising Arts*, July 9, 1930, p. 17.

5. Teague, "The Cash Value of Design," *Arts and Decoration* 43 (Jan. 1936), 48.

6. Loewy, "Designing Towards Profits," *Steel*, Aug. 1, 1938, p. 36.

7. Autobiography, Jamaica version, May 28, 1955, ch. 68, *Geddes*.

8. John W. Higgins, "Design for Mass Production," *The American Magazine of Art* 26 (Sept. 1933), 426.

9. *Horizons* (Boston: Little, Brown, 1932), p. 4.

10. *Technics and Civilization* (New York: Harcourt, Brace, 1934), pp. 211–12, 258, 429–30.

11. "Why Disguise Your Product?" *Electrical Manufacturing*, Oct. 1938, p. 90; Teague, "Designing for Machines," *Advertising Arts*,

April 2, 1930, pp. 19, 21.

12. Teague, "Industrial Art and Its Future," *Art and Industry* 22 (May 1937), 195.

13. "Designing for Machines," p. 21.

14. Philip N. Youtz, "Art and Industry," *American Magazine of Art* 27 (Aug. 1934), 430.

15. Mumford, "Modernism For Sale," *The American Mercury* 16 (April 1929), 455; Teague, "Will It Last?" *Advertising Arts*, March 1931, pp. 17, 19.

16. Dreyfuss, "Everyday Beauty," *House Beautiful* 74 (Nov. 1933), 192; Dreyfuss, "What Is Wrong with American Industrial Designers," lecture delivered May 1, 1934, 1973.15.22(c), *Dreyfuss*; Loewy, "Co-operation Means Successful Designs," *Electrical Manufacturing* 18 (Oct. 1936), 68; and Geddes, *Horizons*, p. 186.

17. Warner, "Modern Form in Design," *Product Engineering* 2 (Aug. 1931), 349; Geddes, *Horizons*, pp. 17–18; and Teague, "Will It Last?" p. 15.

18. Dreyfuss, "Notes for Boston Lecture," March 14, 1933, 1973.15.22(a), *Dreyfuss*; Geddes, "Modern Design," autobiography, Jamaica version, ch. 41–43, *Geddes*.

19. "Modern Design," *Geddes*.

20. Geddes, "The Artist in Industry: An Unrepentant Confession by a Famous Scene Designer," *Theatre Guild Magazine*, Jan. 1930, p. 24.

21. "Modern Design," *Geddes*.

22. *Horizons*, p. 18.

23. Brill, "Our Homesick Plastics," *Plastics and Molded Products* 8 (June 1932), 251, 235; Wright, "On Musical Instruments," *Advertising Arts*, Jan. 1934, p. 21; Teague, "How Much Does Design Stimulate Machine Play?" *The Coin Machine Journal*, Oct. 1933, p. 38.

24. Müller-Munk, "Industrial Design," *Design* 38 (Jan. 1937), 13; Dohner, "Modern Technique of Designing," *Modern Plastics* 14 (March 1937), 71; and Geddes, *Horizons*, p. 223.

25. Geddes, "The Challenge of Industrial Design," autobiography, Jamaica version, ch. 69, 70, *Geddes*; Sheldon Cheney and Martha Candler Cheney, *Art and the Machine: An Account of Industrial Design in 20th-Century America* (New York: Whittlesey House, 1936), pp. 14–15.

26. Teague, "Designing for Machines," p. 23.

27. Teague, "Why Disguise Your Product?" pp. 111–12.

28. Teague, "Rightness Sells," *Advertising Arts*, Jan. 1934, p. 25.

29. Teague, "Machine Age Aesthetics," *Advertising Arts*, July 1932, p. 8.

30. Teague, "Art of the Machine Age," *Industrial Education Magazine* 38 (Nov. 1936), 228.

31. Teague, "Basic Principles of Body Design Arise from Universal Rules," *Society of Automotive Engineers Journal* 35 (Sept. 1934), supp. 18.

32. Teague, "Why Disguise Your Product?" p. 47.

33. Teague, "Industrial Art? Well, Not Exactly," *Commercial Art and Industry* 18 (April 1935), 152.

34. Teague, "Rightness Sells," p. 25.

35. Teague, "Basic Principles of Body Design," supp. 18–19.

36. Teague, "Industrial Art? Well, Not Exactly," p. 152.

37. Teague, "Industrial Art and Its Future," p. 196. Teague probably derived the general drift of his social thought from Mumford. He never credited Mumford, however, possibly because the latter referred to "the studious botching of the kodak" as an example of commercial "stylicizing." Teague in turn ridiculed Mumford's presumed lack of common sense and found *Technics and Civilization* "ponderous, pedantic, but interesting if one can stand the voice of God in which Mr. Mumford habitually speaks." None the less, commercial designer and purist critic shared the same vision. See *Technics and Civilization*, p. 352; "Industrial Art? Well, Not Exactly," pp. 150–53; and Teague, *Design This Day: The Technique of Order in the Machine Age* (New York: Harcourt, Brace, 1940), p. 283.

38. Geddes, *Horizons*, pp. 4–5, 289.

39. *Ladies' Home Journal* 48 (Jan. 1931), 3.

40. Dreyfuss, "Everyday Beauty," p. 190. Emphasis mine.

41. Richard F. Bach, "Toward the Style of the Century," *Advertising Arts*, Jan. 1935, pp. 8–11; Paul T. Frankl, *Machine-Made Leisure* (New York: Harper, 1932), p. 95.

42. Walter Baunard, "The Future Car: How Car Bodies Have Developed at Home and Abroad," *Scientific American*, January 11, 1913, pp. 28–29.

43. Theodore von Kármán, *Aerodynamics: Selected Topics in the Light of Their Historical Development* (Ithaca, N.Y.: Cornell University Press, 1954), p. 9.

44. Raymond Loewy provided an illustration (plate 6) in his essay *The Locomotive: Its Esthetics* (New York: The Studio, 1937).

45. Donald J. Bush, *The Streamlined Decade* (New York: George Braziller, 1975), pp. 55–57.

46. Norman Bel Geddes, "Streamlining," *The Atlantic Monthly* 154 (November 1934), 554–55.

47. Aston, "Body Design and Wind Resistance," *The Autocar*, Aug. 26, 1911, p. 364.

48. See Theodore von Kármán with Lee Edson, *The Wind and Beyond: Theodore von Kármán* (Boston: Little, Brown, 1967), pp. 73, 118, 120, 145; Karl Ludvigsen, "Automobile Aerodynamics: Form and Fashion," *Automobile Quarterly* 6 (Fall 1967), 147.

49. On the Jaray car see Othmar K. Marti, "Streamlining Applied to Automobiles," *Society of Automotive Engineers Journal* 29 (Aug. 1931), 126–27.

50. "Ultra-Streamlined Car Attains Higher Speeds Than Stripped Chassis," *Automotive Industries*, Sept. 27, 1930, pp. 450–51. See also Sir Dennistoun Burney, "The Development of a Rear-Engined Streamline Car," *Society of Automotive Engineers Journal* 30 (Feb. 1932), 57–64; (March 1932), 116–19.

51. Norman G. Shidle, "From the Annual S.A.E. Meeting at Detroit Come Many New Ideas," *Automotive Industries* Jan. 31, 1931, pp. 147–48.

52. Towle, "Projecting the Automobile into the Future," *Society of Automotive Engineers Journal* 29 (July 1931), 33–35, 39.

53. Lay, "Is 50 Miles Per Gallon Possible with Correct Streamlining?" *Society of Automotive Engineers Journal* 32 (April 1933), 144–56; (May 1933), 177–86; 33 (Aug. 1933), 261–67; Heald, "Aerodynamic Characteristics of Automobile Bodies," *Bureau of Standards Journal of Research* 11 (Aug. 1933), 285–91.

54. Fishleigh, "The Tear-Drop Car," *Society of Automotive Engineers Journal* 29 (Nov. 1931), 353–62; Tietjens, "Economy of Streamlining the Automobile," *Society of Automotive Engineers Journal* 30 (March 1932), 150–52.

55. Autobiography, Jamaica version, March 19, 1956, ch. 68, *Geddes*.

56. Le Corbusier, *Towards a New Architecture*, trans. Frederick Etchells (1927; rpt., New York: Frederick A. Praeger, 1963), pp. 121–38. Attacking seasonal changes, Geddes insisted that designers should aim for "the ultimate motor car." See *Horizons*, p. 48.

57. Autobiography, ch. 74; autobiography, Jamaica version, ch. 71, 80–83; *Geddes*.

58. *Horizons*, p. 24. For an engineer's typical interpretation of streamlining's significance see Felix W. Pawlowski, "Wind Resistance of Automobiles," *Society of Automotive Engineers Journal* 27 (July 1930), 5.

59. *Horizons*, p. 293.

60. Mumford, "The Second Wave," *The New Republic*, May 17, 1933, p. 26; Wright, "On Popular Mechanics," *The Saturday Review of Literature*, Dec. 31, 1932, p. 351; and Haskell, "A 'Stylist's' Prospectus," *Creative Art* 12 (Feb.

1933), 126, 132–33.

61. *Industrial Design*, pp. 147–48.

62. Letter from Carl Otto to the author, Dec. 7, 1976.

63. According to a memo from Geddes to Earl Newsom, Oct. 26, 1933, describing a visit to the Chrysler plant, autobiography, Jamaica version, ch. 74, 75, *Geddes*.

64. Michael Lamm, "Magnificent Turkey," *Special-Interest Autos* 4 (April–May 1973), 14. On the Airflow see also "There Are No Automobiles," *Fortune* 2 (Oct. 1930), 73–77; Athel F. Denham, "The 'Airflow' DeSoto," *Automotive Industries*, Dec. 23, 1933, pp. 752–56; "How the Airflows Were Designed," *Automotive Industries*, June 23, 1934, pp. 766-67, 769; Chrysler Corporation Engineering Office, Technical Information, *Story of the Airflow Cars (1934–1937)* (Detroit: Chrysler, 1963); and Beverly Rae Kimes, "Chrysler: From the Airflow," *Automotive Quarterly* 7 (Fall 1968), 206–208.

65. Klemin, "How Research Makes Possible the Modern Motor Car," *Scientific American* 151 (Aug. 1934), 62.

66. *The Saturday Evening Post*, Dec. 16, 1933, p. 31. Ellipses in original.

67. Minutes of meeting, Oct. 19, 1933, file 271, *Geddes*.

68. Minutes, Nov. 15, 1933, file 271, *Geddes*.

69. Minutes, Jan. 10, 1934, file 271, *Geddes*.

70. Chrysler Corporation, *Story of the Airflow Cars*, pp. 9–10.

71. Bush, *The Streamlined Decade*, p. 122.

72. Carl Otto recalled the turmoil at GM in his letter to the author of Dec. 7, 1976.

73. Loewy, "The Evolution of the Motor Car," *Advertising Arts*, March 1934, p. 39.

74. Athel F. Denham, "Engineers Argue Streamlining at S.A.E. Summer Convention," *Automotive Industries*, June 30, 1934, p. 810.

75. In 1935 Stout unveiled such a car, the Scarab, but his plans for commercial production never materialized. See Rich Taylor, "The Prophet: 1935 Stout Scarab," *Special-Interest Autos*, no. 32, Jan.–Feb. 1976, pp. 30, 33.

76. Denham, "Engineers Argue Streamlining," p. 810.

77. Cret to Geddes, Dec. 11, 1934, file 296, *Geddes*.

78. "Record Crowd at SAE Meeting," *Automotive Industries*, June 23, 1934, p. 755; "Discussion at Summer Meeting in Digest," *Society of Automotive Engineers Journal* 35 (Aug. 1934), supp. 27.

79. Page proofs from the 1935 guide, box 59, *Arens*.

80. A printed letter from Alfred P. Sloan, Jr., to GM stockholders, Dec. 11, 1934, box 59, *Arens*.

81. Kettering, "The Motor Car of the Future," *Automobile Topics*, Jan. 26, 1935, p. 763.

82. A phrase used without irony by Daniel C. Sayre, "Streamline," *The American Mercury* 30 (Oct. 1933), 197.

83. Egmont Arens, "Next Year's Cars," *The American Magazine of Art* 29 (Nov. 1936), 736.

84. "Industrial Design Comes of Age," *Business Week*, May 23, 1936, p. 17; "Transport Steamer Princess Anne," *Marine Engineering* 41 (Aug. 1936), 434.

8. Depression into Expression

1. Haskell, "Mixed Metaphors at Chicago," *The Architectural Review* 74 (Aug. 1933), 47–48.

2. Corbett, "The Significance of the Exposition," *The Architectural Forum* 59 (July 1933), 1.

3. "The Streamliners—After Six Years," *Business Week*, Feb. 10, 1940, p. 47.

4. See Ralph Budd, "Railroading Moves Ahead," *The Atlantic Monthly* 155 (May 1935), 534–36; U.S. Bureau of the Census, *Historical Statistics of the United States: Colonial Times to 1957* (Washington, D.C.: Government Printing Office, 1960), p. 430; and John F. Stover, *American Railroads* (Chicago: University of Chicago Press, 1961), pp. 212–17.

5. Arens, "The Train of Tomorrow," *Advertising Arts*, July 1931, pp. 28–29. A list of recipients is in box 27, *Arens*.

6. See "R.R. Futurity," *Business Week*, June 10, 1933, p. 15; "Round and Light," *Fortune* 8 (Aug. 1933), 66; and "Both Fish and Fowl," *Fortune* 9 (Feb. 1934), 94.

7. Railroads soon dropped articulation because changing cars on a train required use of a crane.

8. On development of the UP's *M–10,000* see "Round and Light," pp. 64–66; E. E. Adams, "Light-Weight, High-Speed Passenger Trains," *Mechanical Engineering* 55 (Dec. 1933), 735–40; "Union Pacific Installs Light-Weight High-Speed Passenger Train," *Railway Age*, Feb. 3, 1934, pp. 184–96; "First Streamlined Train," *Business Week*, Feb. 17, 1934, p. 12; and John B. Rae, *Climb to Greatness: The American Aircraft Industry, 1920–1960* (Cambridge: Massachusetts Institute of Technology Press, 1968), pp. 13–14. On the Burlington *Zephyr* see "Burlington 'Zephyr' Completed at Budd Plant," *Railway Age*, April 14, 1934, pp. 533–44; "Wings for the

'Iron Horse,'" *Popular Mechanics* 62 (Aug. 1934), 170–73, 135A; F. G. Gurley, "C. B. & Q. Experiences with Diesel Equipment," *Diesel Power* 14 (May 1936), 294–99; "Pioneer without Profit," *Fortune* 15 (Feb. 1937), 82–87, 128, 130, 132, 134; Alfred P. Sloan, Jr., *My Years with General Motors* (Garden City, N.Y.: Doubleday, 1964), pp. 341–53; and Richard C. Overton, *Burlington Route: A History of the Burlington Lines* (New York: Alfred A. Knopf, 1965), pp. 379–81.

9. "Union Pacific to Test Fast New Aluminum Trains," *News-Week*, June 3, 1933, p. 23.

10. Budd, "Railroading Moves Ahead," p. 538.

11. Cret, "Streamlined Trains," *The Magazine of Art* 30 (Jan. 1937), 18.

12. Stout (as told to Julian Leggett), "Air-Minded Railroading," *Popular Mechanics* 61 (Feb. 1934), 170.

13. See Garet Garrett, "The Articles of Progress," *The Saturday Evening Post*, July 28, 1934, p. 5; Carl R. Gray, "The Effect of Streamlined Trains on Railroading," *Scientific American* 151 (Dec. 1934), 320; Budd, "Railroading Moves Ahead," p. 541; Charles F. A. Mann, "Full Stream Ahead!" *Collier's*, Jan. 11, 1936, p. 13; and Overton, *Burlington Route*, pp. 396–98.

14. For numbers and types of streamliners produced see Mann, "Full Stream Ahead!" p. 42; Franklin M. Reck, *On Time: The History of Electro-Motive Division of General Motors Corporation* (La Grange, Ill.: Electro-Motive, 1948), passim, pp. 87–159. For illustrations of locomotives and trains of the custom-design period see Loewy, *The Locomotive: Its Esthetics* (New York: The Studio, 1937); Kuhler, *My Iron Journey: An Autobiography of a Life with Steam and Steel* (Denver: Intermountain Chapter of the National Railway Historical Society, 1967); Eric H. Archer, *Streamlined Steam* (New York: Quadrant, 1972); Donald J. Bush, *The Streamlined Decade* (New York: George Braziller, 1975), pp. 67–92.

15. Garrett, "The Articles of Progress," p. 5.

16. On streamliner films see Denis Gifford, *Science Fiction Film* (New York: Studio Vista/Dutton, 1971), p. 35.

17. "Streamlined Toys," *Fortune* 11 (Jan. 1935), 16, 20; "Designing to Sell," *Sales Management*, June 15, 1934, p. 583; and "Designing to Sell," *Sales Management*, Oct. 20, 1934, p. 405.

18. "Standard Gas Equipment Co." *Retailing*, Jan. 16, 1933, clipping, 1932–1933 scrapbook, *Geddes*; Antoinette Donnelly, "Streamline the

Silhouette," *New York Sunday News*, Sept. 8, 1935, p. 75, clipping, box 59, *Arens*.

19. Undated advertisement, box 59, *Arens*.

20. See file 478, *Geddes*.

21. *New York Herald Tribune*, April 29, 1934, section 3, p. 5, clipping, box 59, *Arens*.

22. William J. Acker, "Design for Business," *Design* 40 (Nov. 1938), 12.

23. Arens to Chrysler, Jan. 20, 1934, and Jan. 22, 1934, box 14, *Arens*.

24. From a proposal submitted by Arens to the Industries Sales Committee, Nov. 23, 1934, box 19, *Arens*.

25. Arens to Kieth [*sic*] Morgan, Nov. 27, 1934, box 27, *Arens*.

26. Draft of a telegram from Arens to Roosevelt, Nov. 14, 1934, box 27, *Arens*. Secretary M. H. McIntyre acknowledged receipt in a letter to Arens of Nov. 15, 1934.

27. Arens to Kenneth Moore, Dec. 21, 1934, box 19, *Arens*.

28. Arens's report to Industries Sales Committee, Nov. 23, 1934, box 19, *Arens*.

29. Arens to Morgan, Nov. 27, 1934, box 27, *Arens*.

30. Loewy as paraphrased in "Streamlining—It's Changing the Look of Everything," *Creative Design* 1 (Spring 1935), 22.

31. "Streamline Idea in a Commercial Vehicle," *New York Times*, May 6, 1934, section VIII, p. 10.

32. Bush, *The Streamlined Decade*, pp. 36–37.

33. "Robert Heller: American Industrial Designer," *Art and Industry* 25 (Aug. 1938), 69.

34. K. O. Tooker and F. L. Pierce, "The Hoover One Fifty," *Modern Plastics* 14 (Nov. 1936), 32–33, 62–63.

35. The scale is on p. 565 of a Sears catalogue issued from Dallas for fall/winter 1935–1936.

36. Van Doren, *Industrial Design: A Practical Guide* (New York: McGraw-Hill, 1940), pp. 137, 145.

37. Allen, *Since Yesterday* (1939; rpt., New York: Bantam, 1961), p. 2.

38. Vassos, "'Have Designers Missed a Bet?'" *Creative Design* 3 (Oct.–Nov. 1938), 18.

39. See, for example, Katharine Morrison Kahle, *Modern French Decoration* (New York: G. P. Putnam, 1930), pp. 212–13; Russel Wright, "'—Because It Is an Honest Expression of Present-Day Living, Modern Design Should Interest All Thinking Americans,'" *The American Home* 11 (Jan. 1934), 60–62; Earnest Elmo Calkins, "Whither Industrial Design?" *Advertising Arts*, May 1934, p. 10; and Felix Payant, "The Editor's Page," *Design* 37 (March 1936), 1.

40. Geddes might titillate readers of *Ladies' Home Journal* with the curved concrete lines of his "House of Tomorrow," 48 (April 1931), 12–13, 162, but magazines dedicated to homes and domestic interiors devoted most of their space to traditional styles. The "Town of Tomorrow" erected at the New York World's Fair of 1939 by national suppliers of construction materials did not reflect the streamlined architecture of its surroundings. Traditional homes equipped with the latest appliances outnumbered vaguely modern ones by four to one.

41. Hal Foust, "Professor Tries Stream Lines for Buildings," *Chicago Daily Tribune*, Nov. 22, 1934, clipping, box 59, *Arens*. The following discussion has been influenced by two perceptive articles on streamlining and architecture: David Gebhard, "The Moderne in the U.S. 1920–1941," *Architectural Association Quarterly* 2 (July 1970), 4–20; Kathleen Church Plummer, "The Streamlined Moderne," *Art in America* 62 (Jan.–Feb. 1974), 46–54.

42. Architectural historians have noted the similarity of Saarinen's Dulles airport terminal (1958–1963) to one of Mendelsohn's visionary sketches of 1917. With the TWA terminal at Kennedy airport (1959–1962) and the Gateway Arch at St. Louis (completed in the late sixties), it can be seen as a culminating expression of streamlined architecture. All three would have seemed startling but not out of place in the thirties.

43. See Jordy, *American Buildings and Their Architects: The Impact of European Modernism in the Mid-Twentieth Century* (Garden City, N.Y.: Doubleday, 1972), pp. 87–164.

44. See "Albert Kahn," *The Architectural Forum* 69 (Aug. 1938), especially 97–101, 108.

45. "California Gold Rush County Streamlines Its Courthouse," *Architectural Record* 84 (July 1938), 46–48.

46. See "Twentieth Century Limited," *Modern Plastics* 15 (July 1938), 21–23, 62, 64; "New Trains," *The Architectural Forum* 69 (Sept. 1938), 175–82; "Henry Dreyfuss Designs New 'Century' Train," *Design* 40 (Feb. 1939), 5–7; and Lucius Beebe, *20th Century* (Berkeley: Howell-North, 1962), pp. 51–53, 158–66. Both Plummer, "Streamlined Moderne" (p. 48), and Bush, *Streamlined Decade* (pp. 151–53), have discussed the utopian vision of enclosed environments, the latter concluding that approximations "provided psychic insulation from the sight of bread lines, riots and the streets of Hooverville."

47. Johnson, *Machine Art* (New York: Museum of Modern Art, 1934), n. p.

48. Bauer, "Machine-Made," *The American Magazine of Art* 27 (May 1934), 270.

49. "Beauty of Form in Machine Art," *Design* 35 (April 1934), 9.

50. *Art in Our Time* (New York: Museum of Modern Art, 1939), pp. 332–34.

51. Mumford, *Technics and Civilization* (New York: Harcourt, Brace), p. 253.

52. Barr to Geddes, Dec. 4, 1934, file 296, *Geddes*.

53. McAndrew, "'Modernistic' and 'Streamlined,'" *The Bulletin of the Museum of Modern Art* 5 (Dec. 1938), 2.

54. Hitchcock, "Some American Interiors in the Modern Style," *The Architectural Record* 64 (Sept. 1928), 238.

55. "Wins $500 Art Prize," *New York Times*, Oct. 6, 1928, p. 21.

56. Blanche Naylor, "The Designer and Industry," *Design* 34 (Sept. 1932), 83.

57. *Advertising Arts*, Jan. 1934, p. 48.

58. "Art and Machines," *The Architectural Forum* 60 (May 1934), 331. Ironically, the major industrial designers boycotted the National Alliance exhibit of 1935 because they found the organization's aesthetic standards lacking. See "Art: When Manufacturers Run the Show, Designers Won't Play," *News-Week*, April 27, 1935, p. 24; "The Annual Industrial Arts Exposition," *Modern Plastics* 12 (May 1935), 22–23.

59. Moholy-Nagy, "Design Possibilities" (1944), in *Moholy-Nagy*, ed. Richard Kostelanetz (New York: Praeger, 1970), p. 85.

60. D. C. O'Connell, "Industrial Design: Mechanical Technology of Metals," *Architecture* 73 (Jan. 1936), 20.

61. On plastics and streamlined design see Franklin E. Brill, "What Shapes for Phenolics," *Modern Plastics* 13 (Sept. 1935), 21; Brill and Joseph Federico, "Decorative Treatments for Molded Plastics," *Product Engineering* 8 (Jan. 1937), 25; Frank H. Johnson, "Designing Plastic Parts," *Product Engineering* 9 (Feb. 1938), 61; and Raymond P. Calt, "A New Design for Industry," *The Atlantic Monthly* 164 (Oct. 1939), 541–42.

62. Sheldon Cheney and Martha Candler Cheney, *Art and the Machine: An Account of Industrial Design in 20th-Century America* (New York: Whittlesey House, 1936), pp. 217, 291, 98, 102.

63. Teague, "Plastics and Design," *The Architectural Forum* 72 (Feb. 1940), 93–94.

64. Pevsner, *Pioneers of Modern Design: From William Morris to Walter Gropius* (Balti-

more: Penguin, 1968), p. 210.

65. Onderdonk, *The Ferro-Concrete Style: Reinforced Concrete in Modern Architecture* (New York: Architectural Book Publishing Co., 1928), pp. 221, 195.

66. Bragdon, *The Frozen Fountain: Being Essays on Architecture and the Art of Design in Space* (New York: Alfred A. Knopf, 1932), pp. 11–12.

67. Ward, "Towards a New Era of Speed," *Travel* 62 (April 1934), 9–10, 60.

68. Pommer, "Loewy and the Industrial Skin Game," *Art in America* 64 (March–April 1976), 46.

69. *The Streamlined Decade*, p. 4.

70. Thompson as quoted by Bush, *Streamlined Decade*, pp. 9–10.

71. Geddes, *Horizons* (Boston: Little, Brown, 1932), p. 45.

72. Haskell, "From Automobile to Road-Plane," *Harper's Monthly Magazine* 169 (July 1934), 173.

73. Henry Glade, "Future City on Earth" (April 1942), as quoted by Plummer, "Streamlined Moderne," p. 49.

74. Peter Müller-Munk, "The Future of Product Design," *Modern Plastics* 20 (June 1943), 77, 144.

75. Paul T. Frankl, *Machine-Made Leisure* (New York: Harper, 1932), pp. 117, 120.

76. "What Man Has Joined Together . . . ," *Fortune* 13 (March 1936), 71.

77. Burvil Glenn, "A New Definition of Design," *Design* 37 (June 1935), 8.

78. Albert F. Byers, "Eye Appeal in Machinery Design and Finish," *The Iron Age*, May 14, 1936, p. 53.

79. *Art and the Machine*, p. 20.

80. *Technics and Civilization*, p. 357.

81. "Introduction," *The Practice of Design*, ed. Herbert Read (London: Lund Humphries, 1946), p. 21.

9. A Microcosm of the Machine-Age World

1. On the original theme and Teague's influence see "World's Fair, New York Style," *Business Week*, Sept. 28, 1935, p. 18; Donald J. Bush, *The Streamlined Decade* (New York: George Braziller, 1975), p. 159.

2. "Robert (Or-I'll-Resign) Moses," *Fortune* 17 (June 1938), 138. See also Robert A. Caro, *The Power Broker: Robert Moses and the Fall of New York* (New York: Vintage, 1975), pp. 1082–85.

3. *Official Guide Book: New York World's Fair 1939* (New York: Exposition Publications, 1939), pp. 20, 18. Various editions have different paginations, but there is no way to distinguish them.

4. Ibid., p. 27.

5. Ibid.

6. As quoted by R. L. Duffus, "A City of Tomorrow: A New Design of Life," *New York Times Magazine*, Dec. 18, 1938, p. 4.

7. As quoted by Duffus, p. 23.

8. Duffus, p. 4.

9. "San Francisco Golden Gate Exposition 1939," *The Architectural Forum* 70 (June 1939), 464. The comment appears in a comparison of the two fairs.

10. Morton Eustis, "Big Show in Flushing Meadows," *Theatre Arts Monthly* 23 (Aug. 1939), 573.

11. On the theme exhibit consult "Scheme for the Theme Exhibit," Dec. 13, 1938, a spiral notebook containing text, illustrations, and diagrams, 1972.88.258, *Dreyfuss*.

12. Eustis, "Big Show," p. 575.

13. Loewy designed the locomotive specifically as a fair exhibit, according to his letter to Merle Armitage, Dec. 23, 1938, Merle Armitage Collection, Humanities Research Center, University of Texas at Austin.

14. Hamlin, "World's Fairs/1939 Model," *Pencil Points* 19 (Nov. 1938), 676.

15. See clippings in file on Crystal Gazing Palace/TH-13/i.–1., *Geddes*.

16. As quoted by Vivian Vorsanger, "Designers at the Fair," *Printers' Ink Monthly* 35 (Sept. 1937), 22.

17. Teague, "Building the World of Tomorrow: The New York World's Fair," *Art and Industry* 26 (April 1939), 127, 134.

18. Teague, "Industrial Art and Its Future," *Art and Industry* 22 (May 1937), 193.

19. Whalen, "Building the World of Tomorrow," *New York World's Fair 1939* (New York: New York World's Fair, 1936), n. p.

20. Whalen, "What the Fair Means to Business and Industry," *New York World's Fair Bulletin* 1 (June 1937), 1. Emphases mine.

21. Haskell, "To-morrow and the World's Fair," *Architectural Record* 88 (Aug. 1940), 68.

22. Teague, "Exhibition Technique," *American Architect and Architecture* 151 (Sept. 1937), 33.

23. Teague to Fred L. Black of Ford, April 30, 1941, box WDT Sr 3, *Teague*.

24. Teague, "Exhibition Technique," pp. 31–32.

25. Teague to Robert Gregg of U.S. Steel, April 5, 1937, reel 30A, *Teague*.

26. Teague, "Exhibition Technique," pp. 32–34.

27. On these exhibits see *Official Guide*, p. 85; *Science at the New York World's Fair 1939* (mimeographed booklet), p. 10; ibid., p. 18; "Notes on the Design of Chrysler Motors Exhibit," publicity release, *Loewy*; and "New York World's Fair 1939," *The Architectural Forum* 70 (June 1939), 413.

28. Fair publications are described in *Official Guide*, pp. 14–15.

29. Bush discusses the Futurama briefly in *The Streamlined Decade*, pp. 159–63. Another historical account is William Stott's perceptive "Greenbelt and Futurama: The Heavenly Cities of the 1930s," *The Journal of the American Studies Association of Texas* 4 (1973), 18–29. Robert Coombs's "Norman Bel Geddes: Highways and Horizons," *Perspecta 13/14* (1971), pp. 11–27, contains two continuous texts separated by photographs. The upper text is copied verbatim, with some passages omitted but no original material added, without credit, from Geddes, "Description of the General Motors Building and Exhibit to the New York World's Fair," Sept. 8, 1939, a 37-page typescript in file 381, *Geddes*. The lower text incorporates three hundred lines of verbatim material, not set off by quotation marks, from two folders in file 381: "Futurama Conveyor System" and "G M Intersection World's Fair." Despite Coombs's cavalier treatment of scholarly conventions, "his" article provides the fullest published description of the Futurama and the GM building.

30. Haskell, "To-morrow and the World's Fair," p. 71.

31. "Description of the General Motors Building," p. 1; list of building specifications, Aug. 9, 1938; both file 381, *Geddes*.

32. "Description," p. 1, file 381, *Geddes*. Saarinen was credited with forty-six days of work on the project, including the lion's share of scale layouts, sketches, and final drawings for the building's exterior, in "Estimate for General Motors/N.Y. World's Fair Exhibit," prepared by Worthen Paxton, April 27, 1938, file 381, *Geddes*. In a letter to the author, Nov. 5, 1976, Garth Huxtable described Saarinen as "the star designer" on the project but emphasized its collaborative nature.

33. Eustis, "Big Show in Flushing Meadows," p. 571.

34. "Futurama Conveyor System," file 381, *Geddes*.

35. This and following paragraphs are based on "Description"; *Futurama*, a souvenir booklet containing a full text of the recorded narration; other materials in file 381; and films taken from the moving chairs, now part of *Geddes*.

36. John Mason Brown, "Norman Bel Geddes' Addition to the Fair," *New York Post*, May 11, 1939, p. 12.

37. On these borrowings see Stott, "Greenbelt and Futurama," pp. 25–26; Thomas Adams, *The Building of the City: Regional Plan: Volume Two* (New York: Regional Plan of New York, 1931), pp. 306–13, 412–15; Loewy, "The Evolution of the Motor Car," *Advertising Arts,* March 1934, p. 39.

38. Geddes requested ideas on improving traffic flow and lane separation at bridges in a memo to a Mr. MacMurchy of his staff, Sept. 3, 1931, file 16, *Geddes*.

39. "Unfit for Modern Motor Traffic," *Fortune* 14 (Aug. 1936), 94. See also "Miller McClintock," *The National Cyclopaedia of American Biography* 44 (New York: James T. White, 1962), 14–15.

40. From a memo of a meeting of McClintock with Geddes, Dec. 1, 1936, file 356, *Geddes*.

41. See memoranda of staff meetings on Nov. 12, 1936; Nov. 16, 1936; and Dec. 1, 1936; all file 356, *Geddes*.

42. The ads appeared in *Life* and *The Saturday Evening Post* from July to Nov. 1937. For clippings regarding showings of the film and model see "Record Copy Book/Shell/Clippings," file 356, *Geddes*.

43. See "At the Wheel," *New York Times*, June 6, 1937, section 12, p. 12; McClintock, "Of Things to Come," in *New Horizons in Planning: Proceedings of the National Planning Conference, Held at Detroit, Michigan, June 1–3, 1937* (Chicago: American Society of Planning Officials, 1937), pp. 34–38.

44. Bruce Bliven, Jr., "Metropolis: 1960 Style," *The New Republic*, Sept. 29, 1937, p. 212.

45. The contract, in file 381, *Geddes*, was signed on May 3, 1938. Worthen Paxton's "Estimate for General Motors/N.Y. World's Fair Exhibit," file 381, *Geddes*, which contained a detailed breakdown of work already completed on the project, was dated April 27, 1938. Former staffer Garth Huxtable, in a letter to the author, Nov. 5, 1976, described how the staff hastily substituted the name of General Motors for that of Goodyear on all the plans and sketches.

46. See Bradford C. Snell, "American Ground Transport," in *The Industrial Reorganization*

Act: Hearings before the Subcommittee on Antitrust and Monopoly of the Committee on the Judiciary United States Senate . . . on S. 1167 (Washington, D.C.: Government Printing Office, 1974), part 4A, p. A–44. Geddes included this benefit in his sales pitch to GM, but it was dropped from an otherwise identical list of goals announced to the public by GM. See "General Motors Presentation that closed the deal," file 384, *Geddes*; GM press release, July 20, 1938, file 381, *Geddes*.

47. Snell, pp. A–28 to A–31. GM refuted other charges leveled by Snell but failed with this one. See "The Truth about 'American Ground Transport,'" pp. A–112 to A–124.

48. See notes compiled later, in "G M Intersection World's Fair," file 381, *Geddes*.

49. Geddes to John D. Williams, Jan. 28, 1947, autobiography, Jamaica version, ch. 23–32, *Geddes*.

50. Geddes to Frances Waite Geddes, Dec. 13, 1938, in file "Personal Correspondence between Norman Bel Geddes and Frances"; Frances Waite Geddes to Geddes, Dec. 14, 1938, in file "Letters from Frances Waite Geddes to Norman Bel Geddes"; both *Geddes*.

51. Minutes of a meeting, Jan. 27, 1938, file 381, *Geddes*.

52. According to a letter from Geddes to Henry Waite, March 28, 1939, autobiography, ch. 79, *Geddes*. There is no reason to think Geddes's

presentation a determining factor in Roosevelt's appointment two years later of an Interregional Highway Committee—whose purpose was to plan a national highway system to relieve both unemployment and traffic congestion in the postwar period. See Mark Howard Rose, "Express Highway Politics, 1939–1956" (Ph.D. diss., Ohio State University, 1973), p. 61.

53. According to a memo from "L. W." to Geddes and Worthen Paxton, June 16, 1937, and a letter from Geddes to Moses, July 8, 1937, both in file 356, *Geddes*.

54. See "Moses Envisages Future Highways," *New York Times*, Jan. 21, 1940, section 1, p. 11; "Moses Calls Bel Geddes's Plan of City-Shunning Roads 'Bunk,'" *New York Herald Tribune*, Jan. 21, 1940, clipping in file 397, *Geddes*; and "Super-Highways," *New York Times*, Jan. 28, 1940, section 4, p. 8.

55. "Fair's Theme Song Has Its Premiere," *New York Times*, Feb. 3, 1940, p. 9. The full text of the release is in file 397, *Geddes*.

56. Moses to Geddes, March 18, 1940, file 384, *Geddes*.

57. File 409, *Geddes*.

58. "Presentation Plan," April 24, 1940, file 381, *Geddes*. In fact the diorama was broken up.

59. Geddes, *Magic Motorways* (New York: Random House, 1940).

Because most sources for this study are cited in the reference notes to the text I therefore am not including here all books and articles consulted, nor all those cited, nor even all those mentioned in the text. My intention is twofold: to provide readers with a general knowledge of the kinds of material on which my study is based and to point out secondary sources useful in following up subjects that I discuss in passing.

Only within the past fifteen years have historians become interested in industrial design as a business practice and in its products as forms of cultural expression. Most previous treatments resembled Don Wallance's *Shaping America's Products* (New York: Reinhold, 1956), which provides case studies of various product designs in the fifties. Jay Doblin's more recent *One Hundred Great Product Designs* (New York: Van Nostrand Reinhold, 1970) follows the same approach. The first historical study was Jerry Streichler's "Consultant Industrial Designer in American Industry from 1927 to 1960" (Ph.D. diss., New York University, 1962), which traces institutionalization of the profession and discusses contemporary design methods. Streamlining as an industrial design style has been discussed objectively by David Gebhard in "The Moderne in the U.S. 1920–1941," *Architectural Association Quarterly* 2 (July 1970), 4–20, and by Kathleen Church Plummer in "The Streamlined Moderne," *Art in America* 62 (Jan.–Feb. 1974), 46–54. Hostile treatments are by Justin De Syllas, "Streamform: Images of Speed and Greed from the 'Thirties," *Architectural Association Quarterly* 1 (April 1969), 32–41, and Richard Pommer, "Loewy and the Industrial Skin Game," *Art in America* 64 (March–April 1976), 46–47. Martin Greif's *Depression Modern: The Thirties Style in America* (New York: University Books, 1975) contains a short impressionistic essay and numerous photographs—primarily architectural. The first major study of streamlining was Donald J. Bush's *Streamlined Decade* (New York: George Braziller, 1975), derived from his "Streamlining: Functionalism in American Product Design, 1927–1939" (Ph.D. diss., University of New Mexico, 1973). Bush's study provides an excellent, profusely illustrated catalogue of streamlined vehicles but has certain limitations. In addition to excluding most nontransportation designs, he limits discussion to objects that he finds aesthetically valid. Strictly an art historian, he does not discuss the business side of the profession, which I find cannot be separated from the style.

My own study, which does attempt to relate art, business, and culture, is based primarily on two kinds of material not previously utilized to any extent: the archives of industrial designers and articles on design and designers published in art, business, and trade journals. These materials reveal not only how designers operated but also conflicts between ideals and practices.

I focused on Henry Dreyfuss, Norman Bel Geddes, Raymond Loewy, and Walter Dorwin Teague because they were generally considered the leaders of the profession in the thirties. In addition, their papers are available. I would have included Harold Van Doren, who started the decade as a local Toledo designer and rapidly attained national prominence, but I could not locate his papers.

The Henry Dreyfuss Symbols Archive is housed in the Cooper-Hewitt Museum of Design, a division of the Smithsonian Institution in New York City. Meticulously catalogued and listed, the archive contains the "Brown Book," a list of Dreyfuss's commissions, activities, publications, and speeches; hundreds of photographs of his work; a few presentation books prepared for clients; and a small number of manuscripts and letters. Materials from the fifties and sixties predominate. The archive also contains about one hundred file folders of research material pertaining to symbols of every description, gathered by Dreyfuss in his last decade when he became interested in international symbols in business, industry, traffic control, and other fields. Dreyfuss routinely destroyed outdated office correspondence and memoranda, and microfilm reels of sketches and plans have been lost.

The Norman Bel Geddes Collection, the most extensive one I consulted, is in the Hoblitzelle Theatre Arts Library, a division of the Humanities Research Center at the University of Texas at Austin. The collection contains materials from every phase of Geddes's life, ranging from schoolboy drawings and diaries kept by his mother concerning his development as a child, through his short career in advertising and longer one in the theater, to his industrial design period. Included are correspondence (both professional and personal), office memoranda from every period of his life, photographs, sketches, final plans, his personal and professional libraries, source files, models (ranging in scope from a streamlined yacht to a handful of inch-long teardrop cars from the Futurama), films, vacation slides, and even the multitude of rubber stamps with which he routed documents through his office maze. Theatrical materials have been catalogued and listed by the library staff. Industrial design materials fill about twenty file drawers. They are still uncatalogued, but Geddes's own indexed filing system is adequate. The collection also contains two manuscript versions of Geddes's uncompleted autobiography as well as manuscripts of his other books.

Raymond Loewy kindly permitted me to use files of photographs and public relations materials in the office of Raymond Loewy International in New York City. His business papers from the thirties were destroyed in the fifties.

Through the historical interests of Arthur J. Pulos, chairman of the Department of Design at Syracuse University, the George Arents Research Library for Special Collections at Syracuse University has become a center for the study of industrial design. In addition to housing the Walter Dorwin Teague Collection, the library also contains collections devoted to Egmont Arens (89 archival boxes), Raymond Spilman (25 boxes), Brooks Stevens (one box), John Vassos (one box), and Russel Wright (79 boxes). Few of these materials originate from the thirties.

The Teague Collection contains two packing cartons with materials from the designer's boyhood and advertising career, both manuscripts and examples of his early work. Fifty-seven reels of microfilm contain client correspondence, office memoranda, indexed lists of sketches and drawings, and a very few sketches and photographs, dating from about 1932 to 1960. About sixty archival boxes contain source clippings collected by Teague prior to 1931 and office papers relating to projects in the late fifties, including the interior design of the U.S. Air Force Academy. Unfortunately, the collection has no models and virtually no graphic materials relating to Teague's designs.

Books written by the designers provided insight into their careers. Dreyfuss's *Designing for People* (New York: Simon and Schuster, 1955) and Loewy's *Never Leave Well Enough Alone* (New York: Simon and Schuster, 1951) both contain anecdotal biographical material as well as simplified accounts of the designers' methods and philosophies. The published version of Geddes's autobiography, *Miracle in the Evening* (Garden City, N.Y.: Doubleday, 1960), covers only his theatrical career, but *Horizons* (Boston: Little, Brown, 1932), provides information about his first years in industrial design. Teague's *Design This Day: The Technique of Order in the Machine Age* (New York: Harcourt, Brace, 1940), which includes numerous photographs of his work, is the most philosophical account of industrial design. For accessible illustrations see also Dreyfuss's *Ten Years of Industrial Design* (New York: 1939) and Lois Frieman Brand's *Designs of Raymond Loewy* (Washington, D.C.: Smithsonian Institution Press, 1975).

This study could not have been written without the use of hundreds of articles written by and about designers or descriptive of their work. Often blatant propaganda but sometimes serious, such articles reveal relationships between industrial design and both business and the public. They provide case histories, often illustrated, of the design of specific products, vehicles, or buildings. I gained access to this material originally through the usual indexes, *The Reader's Guide to Periodical Literature* and *The New York Times Index*, but quickly found it necessary to consult specialized indexes, such as *The Art Index* and *The Engineering Index*. Most essential was *The Industrial Arts Index*. Frequency of reference to four or five periodicals indicated the necessity of skimming through all their volumes for the years covered by this study. Parenthetically, two essential but neglected biographical reference sets are *Current Biography* and *The National Cyclopaedia of American Biography*, both of which include individuals prominent in their fields but generally unknown to the public even in their own time.

During the thirties industrial design at times dominated two periodicals dedicated to other subjects: *Advertising Arts*, a slick graphics supplement to *Advertising and Selling*, and *Modern Plastics* (earlier known as *Plastics and Molded Products* and *Plastic Products*). Only *Art and Industry* (also called *Commercial Art and Industry*) devoted itself exclusively to industrial design. It was published in Britain, but Americans used it as a forum. *International Studio*, an American edition of Britain's *Studio*, and *Arts and Decoration* provided coverage of the modernistic style and of interior design, while *Design*, primarily a magazine for school art instructors, kept abreast of industrial design developments. Not until the fifties, with the advent of *Industrial Design*, did the profession support its own journal, which sometimes carries short retrospectives and historical essays.

Two business periodicals provided publicity for industrial design, *Business Week* rather noncommittally and *Fortune* enthusiastically (the latter is essential as well for coverage of corporations for which designers worked). Trade journals for fields related to industrial design, such as *Printers' Ink*, *Printers' Ink Monthly*, *Factory and Industrial Management*, *Sales Management*, and, reluctantly, *Product Engineering*, occasionally devoted space to the profession. Other trade journals provided designers space to boost their services and occasionally published success stories. Among these were *The Coin Machine Journal*, *Electrical Manufacturing*, *The Iron Age*, *Metal Progress*, and *Steel*.

Despite the animosity expressed toward industrial designers by some architects, their journals covered transportation interiors, commercial facades, and exposition buildings produced by designers. Most active in this regard were *The Architectural Forum* and *Architectural Record*, but *American Architect and Architecture* also covered the field, as did *Pencil Points*, which published surveys of several designers' work. All are excellent sources for studying the modernistic and streamlined styles in architecture.

Popularization of streamlining can be traced in *Popular Mechanics* and *Popular Science Monthly*. *Scientific American*, not as "respectable" then as now, included occasional articles. For tracing the development of streamlining in transportation, *Railway Age*, *Automotive Industries*, and the *Society of Automotive Engineers Journal* are essential. Historical articles of worth can be found in *Trains* (also called *Trains and Travel*), *Automotive Quarterly*, and *Special-Interest Autos*— all far above the average special-interest magazine in quality.

Geddes and Teague were not alone in writing about their profession in the thirties. Roy Sheldon and Egmont Arens in *Consumer Engineering: A New Technique for Prosperity* (New York: Harper, 1932) provided a manifesto on manipulating consumers through design. Harold Van Doren's *Industrial Design: A Practical Guide* (New York: McGraw-Hill, 1940) is a balanced account of all aspects of the profession, from modeling techniques through methods of billing clients. Sheldon Cheney and Martha Candler Cheney glorified industrial design and its expressionist streamlined style in their profusely illustrated *Art and the Machine: An Account of Industrial Design in 20th-Century America* (New York: Whittlesey House, 1936). Theatrical sources of industrial design can be found in

Oscar G. Brockett's *History of the Theatre* (Boston: Allyn and Bacon, 1968). The major functionalist histories of design and architecture are Siegfried Giedion's *Space, Time and Architecture: The Growth of a New Tradition* (Cambridge: Harvard University Press, 1941) and Nikolaus Pevsner's *Pioneers of Modern Design: From William Morris to Walter Gropius* (Baltimore: Penguin, 1972), originally published in 1936 as *Pioneers of the Modern Movement*. Rudolph Rosenthal and Helena L. Ratzka's *Story of Modern Applied Art* (New York: Harper, 1948) surveys the decorative tradition before industrial design. Bevis Hillier's *Art Deco of the 20s and 30s* (New York: E. P. Dutton, 1968) provides the best introduction to that subject. A sense of the purist opposition to commercial design emerges from Russell Lynes's *Good Old Modern: An Intimate Portrait of The Museum of Modern Art* (New York: Atheneum, 1973). For a description of the cutthroat legal world of designers see Sylvan Gotshal and Alfred Lief, *The Pirates Will Get You: A Story of the Fight for Design Protection* (New York: Columbia University Press, 1945). Among cultural analyses of technology and design the best are Lewis Mumford's *Technics and Civilization* (New York: Harcourt, Brace, 1934), read by most designers in the thirties, Siegfried Giedion's *Mechanization Takes Command: A Contribution to Anonymous History* (New York: W. W. Norton, 1969), a study of cultural effects of mass production, first published in 1948, and John A. Kouwenhoven's *Arts in Modern American Civilization* (New York: W. W. Norton, 1967), first published in 1948 as *Made in America*, which postulates an American technological vernacular style of machine design.

A survey of American architecture is John Burchard and Albert Bush-Brown's *Architecture of America: A Social and Cultural History* (Boston: Little, Brown, 1966). Carl W. Condit's *American Building Art: The Twentieth Century* (New York: Oxford University Press, 1961) discusses technical developments conducive to streamlined architecture. From Reyner Banham's *Theory and Design in the First Machine Age* (New York: Frederick A. Praeger, 1967), first published in 1960, a comparison of architectural rhetoric and practice, one gathers that functionalists were not as "pure" as they thought. Revisionist accounts of modern architecture are Vincent Scully, Jr.'s *Modern Architecture: The Architecture of Democracy* (New York: George Braziller, 1974, originally published in 1961), Reyner Banham's *Guide to Modern Architecture* (London: The Architectural Press, 1962), and William H. Jordy's superb *American Buildings and Their Architects:*

The Impact of European Modernism in the Mid-Twentieth Century (Garden City, N.Y.: Doubleday, 1972). For illustrations of buildings at the major American expositions of the thirties see Arnold L. Lehman's *1930's Expositions* (Dallas: Dallas Museum of Fine Arts, 1972).

An earlier efficiency movement, that of production rather than distribution, is explored in Samuel Haber's *Efficiency and Uplift: Scientific Management in the Progressive Era 1890–1920* (Chicago: University of Chicago Press, 1964). The Technocracy movement of the thirties, whose socialist approach to overproduction opposed the consumer engineering solution, is treated by Henry Elsner, Jr., in *The Technocrats: Prophets of Automation* (Syracuse: Syracuse University Press, 1967) and more cogently by William E. Akin in his *Technocracy and the American Dream: The Technocrat Movement, 1900–1941* (Berkeley: University of California Press, 1977).

Daniel J. Boorstin gives a perceptive account of the rise of an American consumer society in his encyclopedic *The Americans: The Democratic Experience* (New York: Random House, 1973). The effect of material affluence on Americans has been traced by David M. Potter in *People of Plenty: Economic Abundance and the American Character* (Chicago: University of Chicago Press, 1954). Excellent sociological studies that provide insight into the life of typical Americans in the twenties and thirties are Robert S. Lynd and Helen Merrell Lynd's *Middletown: A Study in Contemporary American Culture* (New York: Harcourt, Brace, 1929) and *Middletown in Transition: A Study in Cultural Conflicts* (New York: Harcourt, Brace, 1937). Useful summaries of technological developments are found in Melvin Kranzberg and Carroll W. Pursell, Jr., eds., *Technology in Western Civilization* (New York: Oxford University Press, 1967), II, while the sociological and cultural impact of technological developments can be studied through essays compiled by President Hoover's Research Committee on Social Trends, *Recent Social Trends in the United States* (New York: McGraw-Hill, 1933), and statistically through U.S. National Resources Committee, Subcommittee on Technology, *Technological Trends and National Policy: Including the Social Implications of New Inventions* (Washington, D.C.: Government Printing Office, 1937) and U.S. Bureau of the Census, *Historical Statistics of the United States: Colonial Times to 1957* (Washington, D.C.: Government Printing Office, 1960).

This study relies on numerous histories of individual businesses and corporations. Most are listed in either of two excellent bibliographical studies: Henrietta M. Larson's *Guide to Business History* (Cambridge: Harvard University Press, 1948) and Lorna M. Daniells's *Studies in Enterprise: A Selected Bibliography of American and Canadian Company Histories and Biographies of Businessmen* (Boston: Baker Library, Harvard University Graduate School of Business Administration, 1957). On general business history for the period under consideration see Alfred D. Chandler, Jr.'s *Strategy and Structure: Chapters in the History of the American Industrial Enterprise* (Cambridge: The MIT Press, 1962) and Thomas C. Cochran's *American Business in the Twentieth Century* (Cambridge: Harvard University Press, 1972). In *The Visible Hand: The Managerial Revolution in American Business* (Cambridge: Harvard University Press, 1977), Chandler describes the integration of mass production and mass distribution up to about 1920. On the interactions of science and business see David F. Noble's *America by Design: Science, Technology and the Rise of Corporate Capitalism* (New York: Alfred A. Knopf, 1977).

Advertising, the field from which industrial design emerged, is discussed cogently by Otis Pease in *The Responsibilities of American Advertising: Private Control and Public Influence, 1920–1940* (New Haven: Yale University Press, 1958). Neil H. Borden's massive *Economic Effects of Advertising* (Chicago: Richard D. Irwin, 1942) is still indispensable. "Fifty Years: 1888–1938," *Printers' Ink*, July 28, 1938, section 2 (472 pp.), provides a useful index of changing attitudes and practices in advertising. In related fields Albert Q. Maisel's *100 Packaging Case Histories* (New York: Breskin, 1939) gives a graphic survey of the state of the art, while public relations in the twenties and thirties can be studied through Edward L. Bernays, *Biography of an Idea: Memoirs of Public Relations Counsel Edward L. Bernays* (New York: Simon and Schuster, 1965).

John F. Stover's two volumes, *American Railroads* (Chicago: University of Chicago Press, 1961) and *The Life and Decline of the American Railroad* (New York: Oxford University Press, 1970), provide the only general account of the industry, which has yielded a rich volume of corporate histories. Streamliners, both steam and diesel, are profusely illustrated in Lucius Beebe's *Trains in Transition* (New York: D. Appleton-Century, 1941) and Eric H. Archer's *Streamlined Steam* (New York: Quadrant, 1972). The plastics industry, which has so changed the American environment, deserves fuller treatment than provided in J. Harry Dubois's disorganized *Plastics History U.S.A.* (Boston: Cahners, 1972). A technological history of aviation, including some discussion of practical applications of aerodynamics, is Charles Harvard Gibbs-Smith's *Aviation: An Historical Survey from Its Origins to the End of World War II* (London: Her Majesty's Stationery Office, 1970). More industry and business-oriented is John B. Rae's *Climb to Greatness: The American Aircraft Industry, 1920–1960* (Cambridge: The MIT Press, 1968). Rae, primarily a historian of the automobile industry, has written two excellent volumes on the subject: *American Automobile Manufacturers: The First Forty Years* (Philadelphia and New York: Chilton, 1959) and *The Road and the Car in American Life* (Cambridge: The MIT Press, 1971). His *American Automobile: A Brief History* (Chicago: University of Chicago Press, 1965) is primarily a condensa-

tion of the former. City planning, including evolution of traffic control devices and urban expressways, can best be studied through Mel Scott's *American City Planning Since 1890* (Berkeley and Los Angeles: University of California Press, 1969). Political considerations involved in highway planning are surveyed by Mark Howard Rose in his "Express Highway Politics, 1939–1956" (Ph.D. diss., Ohio State University, 1973), which includes a brief survey of developments prior to 1939. Finally, for a description of the state of the art of highway planning and construction when Geddes envisioned the superhighways of 1960, see John H. Bateman, *Introduction to Highway Engineering: A Textbook for Students of Civil Engineering* (New York: John Wiley, 1939).

Numerals in italics indicate illustrations.